T0142989

Advanced Theory and Applications of Engineering
Systems Under the Framework of Industry 4.0

Yongsheng Ma

Editor

Advanced Theory and Applications of Engineering Systems Under the Framework of Industry 4.0

Proceedings of 2022 International Conference
on Intelligent Systems Design
and Engineering Applications

Editor
Yongsheng Ma
Southern University of Science
and Technology
Shenzhen, Guangdong, China

ISBN 978-981-19-9827-0   ISBN 978-981-19-9825-6 (eBook)
https://doi.org/10.1007/978-981-19-9825-6

This Springer imprint is published by the registered company Springer Nature Singapore Pte Ltd.
The registered company address is: 152 Beach Road, #21-01/04 Gateway East, Singapore 189721,
Singapore

Preface

The 2022 International Conference on Intelligent Systems Design and Engineering Applications (ISDEA 2022) was planned in Tokyo, Japan, during May 13–15, 2022. After serious consideration, the committee has made the difficult decision to have it in a fully virtual mode. We were looking forward to seeing everyone in person, but we are excited about the opportunity to innovate by creating an engaging virtual conference that will be rewarding for both presenters and attendees.

The major goal of the conference is to bring together academic scientists, engineers, and industry researchers to share their experiences and research results and discuss the challenges encountered and the solutions adopted. Prestigious experts and professors have been invited to present the latest update in their respective expertise areas.

This year, we have almost 50 submissions from all over the world, and 20 papers are selected for publication by Springer. ISDEA 2022 proceedings span over three created topic tracks including *Advanced Electronics* and *Information Technology, Mobile Robot and Intelligent Control System, and Modern Energy and Electrical Engineering.* As per the program, we had four speeches and two technical sessions on May 14, and we had another four speeches and three technical sessions on May 15. We hope this virtual mode has brought a new level of style to the international conference in such a difficult period.

On behalf of the conference committee, we wish to thank the keynote speakers, invited speakers, and authors of the selected papers for their outstanding contributions. We would also like to thank members of the organizing committee, anonymous reviewers, and volunteers for their great efforts. Without their contribution, dedication, and commitment, we would not have achieved so much. Thanks also go to the Conference Secretaries for their dedicated cooperation. We sincerely hope that you will find ISDEA 2022 beneficial and fruitful for your professional development.

Shenzhen, China   Prof. Yongsheng Ma
ISDEA 2022 Conference Co-Chair

Conference Committees

Conference Chair

Prof. Makoto Iwasaki, Nagoya Institute of Technology, Japan

Conference Co-Chair

Prof. Yongsheng Ma, SUSTech in Shenzhen, China

Conference Program Chairs

Prof. Wilson Q. Wang, Lakehead University, Canada
Prof. LAW, Nancy W. Y., The University of Hong Kong, China
Assoc. Prof. Cheng Siong Chin, Newcastle University, Singapore

Conference Publication Chair

Prof. Nobuo Funabiki, Okayama University, Japan

Conference Program Co-Chair

Assoc. Prof. Xian-Hua Han, Yamaguchi University, Japan

Technical Program Committees

Prof. Francis Richard Ferraro, University of North Dakota, USA
Prof. Juan M. Górriz Sáez, University of Granada, Spain
Prof. Abdulkareem Al-Alwani, Yanbu University College, Saudi Arabia
Prof. Hassan M. H. Mustafa, Al-Baha University K.S.A., Egypt
Prof. Ritthiwut Puwaphat, Princess of Naradhiwas University, Thailand
Prof. Dr. Loc Nguyen, Sunflower Soft Company, Vietnam
Prof. Nakhat Nasreen, Aligarh Muslim University, India
Prof. Gurendra Nath Bhardwaj, NIIT University, India
Assoc. Prof. Caruso Giandomenico, Politecnico di Milano, Italy
Assoc. Prof. Barry Kissane, Murdoch University, Australia
Assoc. Prof. Jeffrey Carruthers, Boston University, USA
Assoc. Prof. Robert G. Hansen, Norwegian Business School, Norway
Assoc. Prof. Tiffany Y. Tang, Wenzhou-Kean University, China
Assoc. Prof. Violeta Kalnytska, Kharkiv National Medical University, Ukraine
Assoc. Prof. Yu-Mei Wang, University of Alabama at Birmingham, USA
Assoc. Prof. Leonardo Garrido, Tecnológico de Monterrey, Campus Monterrey, Mexico
Assoc. Prof. Ahmed Moumena, Hassiba Benbouali University of Chlef, Algeria
Dr. Ahmad Jazlan Haja Mohideen, International Islamic University Malaysia, Malaysia
Dr. Vladislava Sarkis-Ivanova, Kharkiv National Medical University, Ukraine
Dr. Nik Zulkarnaen Khidzir, Universiti Malaysia Kelantan, Malaysia
Dr. Eunice B. Custodio, Bulacan State University, Philippines
Dr. Chau Kien Tsong, Universiti Sains Malaysia, Malaysia
Dr. Kai Ming Kiang, The Chinese University of Hong Kong, China
Dr. Krongthong Khairiree, Suan Sunandha Rajabhat University, Thailand
Dr. Sunghye Lee, Korea Advanced Institute of Science and Technology, South Korea
Dr. Thaweesak Yingthawornsuk, King Mongkut's University of Technology Thonburi, Thailand
Dr. Raymond Li, University of Canberra, Australia
Dr. Siti Hajar Binti Halili, Unversity of Malaya, Malaysia
Dr. P. K. Paul, Raiganj University, India

Contents

Advanced Electronics and Information Technology

A Transformer-Based Network with Character-Level Masks for CAPTCHA Recognition

Ke Qing and Rong Zhang

Abstract Text-based CAPTCHA recognition remains a hot research topic in artificial intelligence and Internet security. The end-to-end deep learning model, which simultaneously localizes and recognizes the character, has recently become the mainstream framework to solve CAPTCHAs. However, prior works are still limited in generalization ability when the training set is relatively small, and the performance achieved on some CAPTCHA schemes remains to be improved. In this paper, we introduce a transformer-based model that uses a novel self-attention mechanism with character-level masks to solve text-based CAPTCHAs. This mechanism enables the encoder and decoder to learn the effective representation of the non-semantic character sequence. Experimental results show that our model has superior generalization ability, even when the training set contains only 600 samples. Meanwhile, our model significantly improves the robustness to different CAPTCHA schemes, achieving a high accuracy above 90% on five representative CAPTCHAs deployed on popular websites.

Keywords Text-based CAPTCHA recognition · Deep neural network · Transformer-based model · Website security

1 Introduction

Text-based CAPTCHA (Completely Automated Public Turing Test to Tell Computers and Humans Apart) is a widely used security mechanism on the Internet [1]. It effectively protects websites from malicious automatic requests, such as registration, comment, and rating. While some researchers have proposed alternative CAPTCHAs schemas, the text-based CAPTCHA maintains the dominance due to its

K. Qing · R. Zhang (✉)
Department of Electronic Engineering and Information Science, University of Science and
Technology of China, Anhui Province, No. 443 Huangshan Rd, Hefei 230027, P. R. China
e-mail: zrong@ustc.edu.cn

K. Qing
e-mail: qingkeld@mail.ustc.edu.cn

© The Author(s), under exclusive license to Springer Nature Singapore Pte Ltd. 2023   3
Y. Ma (ed.), *Advanced Theory and Applications of Engineering Systems Under
the Framework of Industry 4.0*, https://doi.org/10.1007/978-981-19-9825-6_1

flexibility and ease of use [2]. Table 1 shows the schemes of text-based CAPTCHAs deployed on some popular websites.

Meanwhile, breaking text-based CAPTCHAs remains a hot research topic in artificial intelligence and Internet security. The effectiveness of CAPTCHA schemes deployed on popular websites are evaluated by different recognition methods, especially machine learning based attacks. The prior works to solve CAPTCHAs can be categorized as segmentation-based and end-to-end methods.

The segmentation-based method divides the text in the CAPTCHA image into sub-region containing a single character and then recognizes the sub-region with machine learning models, such as support vector machine (SVM) [3] and convolutional neural network (CNN) [4, 5]. The segmentation step dominates the performance of this kind of method since the single character can be accurately recognized once it is apart from the text sequence [6].

On the other hand, the end-to-end method has recently become the mainstream framework for simultaneously localizing and recognizing the character. This method typically uses a CNN to extract features of the CAPTCHA image and a recurrent neural network (RNN) to decode the output sequence from the features [7–9]. However, the prior end-to-end works are limited in generalization ability when the training set is relatively small. For the scheme of CAPTCHA changes frequently, the model that achieves high performance based on extensive labeled data is prohibitive. Meanwhile, the recognition accuracy of the prior model is sensitive to different schemes, and the robustness to security features, such as distortion, overlapping, and background noise remains to be improved.

Inspired by the Transformer model that has achieved superior performance in natural language processing [10–12] and image recognition tasks [13, 14], we

Table 1 The schemes of text-based CAPTCHAs deployed on popular websites

Website	CAPTCHA scheme	Background noise	Varied fonts	Distortion	Overlapping
Google			●	●	●
Baidu			●	●	●
QQ		●			●
Sohu		●			
Facebook		●		●	●

(continued)

Table 1 (continued)

Website	CAPTCHA scheme	Background noise	Varied fonts	Distortion	Overlapping
360		●			
JD.com		●	●		●
Amazon		●	●		●
Sina			●	●	●
eBay			●	●	●
Wikipedia				●	

propose a transformed-based model that uses a novel self-attention mechanism with character-level masks to solve text-based CAPTCHAs. In our model, the encoder focuses on the characters sharing overlapped strokes we refer to as confusing pairs, and the decoder focuses on only the previous one character due to the semantic absence in the text sequence. This mechanism enables our model to learn the effective representation of the CAPTCHA images to improve the generalization ability and robustness.

We evaluate the performance of our model on different sizes of the training set. The experiment shows that our model achieves superior performance even when the training set contains only 600 samples. Furthermore, we use the model to attack five text-based CAPTCHA schemes, which cover representative security features shown in Table 1 and achieve a high accuracy above 90% on all schemes.

The remainder of this paper is organized as follows: Sect. 2 provides an overview of the related CAPTCHA breaking methods. Section 3 presents the details of our transformer-based model and demonstrates the mechanism of character-level masks based on the property of text-based CAPTCHAs. Section 4 evaluates the performance of our model and makes a comprehensive analysis. Finally, Sect. 5 concludes the paper.

2   Related Work

Early text-based CAPTCHA consists of separate Arabic numbers and letters. Conventional optical character recognition (OCR), specifically scene text recognition (STR) [15–17], can easily solve the CAPTCHA. To reduce the vulnerability of the text-based CAPTCHA, researchers introduce security features, such as background noise, distortion, and overlapping, to protect CAPTCHA schemes from attacks based on STR. This section introduces an overview of the prior works to solve CAPTCHAs. We categorize the attacks as segmentation-based and end-to-end frameworks.

2.1   Segmentation-Based Attack

As shown in Fig. 1a, the Segmentation-based attack framework consists of multiple steps to process the input image and predict the text sequence. A typical work-flow contains preprocessing, segmentation, recognition, and post-processing. The performance of this attack depends largely on the effectiveness of the segmentation method.

Yan and Ahmad [18] proposed a low-cost attack based on the analysis of Microsoft CAPTCHA and achieved a success rate higher than 90%. They modified the segmentation method to be appropriate for Yahoo and Google CAPTCHA, respectively, achieving success rates of about 60 and 8.7%. Starostenko et al. [3] proposed a three-color bar code as the segmentation method to break reCAPTCHA (2012). They used an SVM-based learning classifier to recognize the character and broke four CAPTCHAs with success rates ranging from 40.4% to 94.3%. Gao et al. [19] proposed a segmentation method based on the Gabor filter to break CAPTCHA schemes such as reCAPTCHA, Yahoo, and Baidu. They used KNN as the classifier and achieved success rates ranging from 5 to 77%. Gao's team [5] also presented a mixed segmentation method to break a two-layer Microsoft CAPTCHA at a 44.96% success rate.

The segmentation-based attack relies on the pixel-level property of the CAPTCHA scheme. Therefore, the attack applies to one CAPTCHA scheme or a few schemes sharing similar security features. Moreover, the heavy expert involvement and time-consuming process limit the feasibility of segmentation-based attacks.

2.2   End-To-End Attack

The end-to-end neural network has recently become the mainstream framework to solve CAPTCHAs. As shown in Fig. 1b, the model localizes and recognizes the character simultaneously. Street View and reCAPTCHA teams in Google [20] proposed a deep convolutional neural network to recognize the multi-digit numbers from Street

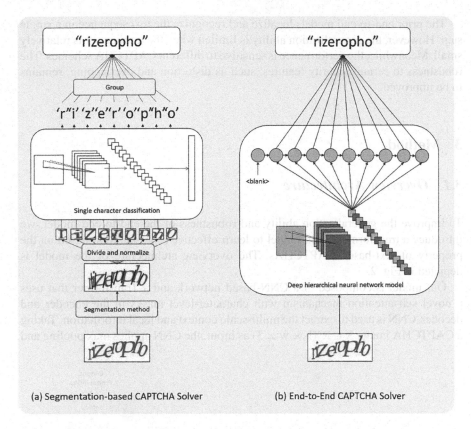

Fig. 1 The framework of the segmentation-based solver and the end-to-end solver

View imagery. They further applied it to transcribing synthetic distorted text from reCAPTCHA and achieved a 99.8% accuracy. This approach integrated localization, segmentation, and recognition steps via CNN. However, the performance reported in this work was taken from a training set in order of millions of CAPTCHA images.

There has also been much interest in utilizing RNN to solve the CAPTCHAs as text sequences. Chen et al. [7] proposed a two-dimensional LSTM decoding method to break their generated CAPTCHAs with a 55.2% accuracy. Shu et al. [8] used a CNN-RNN network consisting of a deep residual convolutional neural network and a two-layer GRU network. They achieved 99.67% accuracy with the 50,000 training samples and 72.9% with 4,000 training samples. The CAPTCHA used in this work is generated, and the security features were relatively simple. Yang Zi and Haichang Gao [9] proposed a model consisting of a CNN and attention-based RNN. This attack broke CAPTCHA schemes deployed by 11 popular websites such as Google, Wikipedia, Baidu, QQ, and Sina, achieving a success rate ranging from 74.8% to 97.3%. The training set size is 200,000 for Google CAPTCHA and 8,500 for the rest of the schemes.

The prior end-to-end models localize and recognize the text sequence in a single step. However, the generalization ability is limited when the training set is relatively small. Meanwhile, the performance is sensitive to different CAPTCHA schemes. The robustness to certain security features, such as distortion and overlapping, remains to be improved.

3   Method

3.1   Overview Architecture

To improve the generalization ability and robustness of the end-to-end model, we introduce a transformer-based model to learn effective representation based on the property of text-based CAPTCHAs. The overview architecture of the model is depicted in Fig. 2.

Our model is composed of a CNN-based network and a Transformer that uses a novel self-attention mechanism with character-level masks in the encoder and decoder. CNN is used to extract the multi-scale context and local information. Taking a CAPTCHA image $X \in R^{(h \times w \times 3)}$ as input, the CNN applies max pooling and

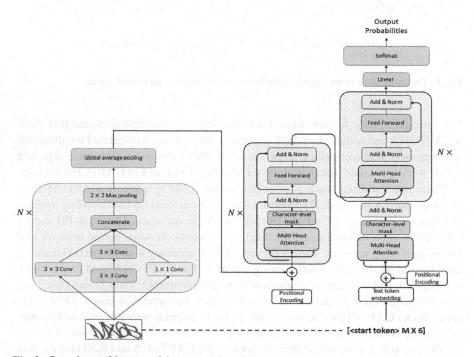

Fig. 2 Overview architecture of the proposed model

Table 2 The detail of convolutional layers in the CNN module

Type	Kernel size	Input size	Number of parameters
Conv	3×3	$40 \times 100 \times 3$	992
Max pooling	2×2	$40 \times 100 \times 32$	–
Mixed-Conv	$1 \times 1, 3 \times 3$	$20 \times 50 \times 32$	9,920
Max pooling	2×2	$20 \times 50 \times 64$	–
Mixed-Conv	$1 \times 1, 3 \times 3, 5 \times 5$	$10 \times 25 \times 64$	60,032
Max pooling	2×1	$10 \times 25 \times 128$	–
Mixed-Conv	$1 \times 1, 3 \times 3, 5 \times 5$	$5 \times 25 \times 128$	165,504

global average pooling to downsample X. Next, the feature sequence $f \in R^{\wedge}(L \times d\_h)$ extracted from the convolutional layers is fed to the Transformer encoder, where L is the sequence length and d_h is feature dimension of the last convolutional layer. In the training stage, right-shifted text sequence is embedded with sinusoid position and then fed to the decoder. Finally, the decoder outputs the prediction probability of size [N_c,C], where N_c is the number of characters in text sequence and C is the number of character classes.

3.2 CNN-Based Feature Extraction

We use a CNN similar to inception-v3 [21] that achieved the superior performance on the ImageNet classification challenge. We performs convolution with 3 different sizes of filters ($1 \times 1, 3 \times 3, 5 \times 5$) to learn the representation of the input CAPTCHA image in multi-scale. Additionally, the outputs are concatenated and sent to the next module. Note that we factorize 5×5 convolution to two 3×3 convolution operations to increase the non-linearity of the model and improve computational speed. The architecture is shown as Table 2.

3.3 Self-attention with Character-Level Mask

The prior end-to-end model employed the RNN to model the sequential text such as LSTM [22], GRU [8], and attention-based RNN [9]. However, the text sequence consisting of pixel-level relative characters in the CAPTCHA is non-semantic and therefore memorizing the long term of dependencies is meaningless. Meanwhile, the neighboring characters sharing overlapped strokes we refer to as confusing pairs restrict the performance of the prior model. For example, the connected 'Q-J' is a typical confusing pair, which is easily misclassified as 'A-U' and 'C-U'.

Table 3 Confusing pairs sharing overlapped strokes in Sina CAPTCHA

Confusing pair	CAPTCHA Image	Model prediction	Confidence	Truth confidence
Q-J		Q-U	0.7776	0.0672
		C-U	0.6279	0.1297
C-H		D-Y	0.8100	0.0650
		O-Y	0.8655	0.1153
N–N		M-I	0.7593	0.0911
		N-V	0.6344	0.1449

Table 3 shows some confusing pairs of Sina CAPTCHA. The first column lists the labeled ground truth of the pairs, and we use a well-trained CNN-RNN model similar to [12] to recognize some unseen CAPTCHA images in the second column. The 'model prediction' column shows the output pairs given by this model at the confidence in the following column. Finally, the last column lists the confidence corresponding to the ground truth. Note that confidence is the product of the predicted probability in the softmax layer.

In this section, we propose a Transformer-based module to learn the effective representation of the neighboring characters that are pixel-level relative. Based on the self-attention mechanism that relates the features in different positions, we introduce character-level masks into the encoder and decoder, forcing self-attention to focus on the neighboring characters.

Encoder. Our model removes the input embedding of the original Transformer. The encoder takes the representation learned from the CNN-based module as input directly. We firstly encode the position information into the feature sequence:

$$f_i = \sqrt{d_h} \cdot f_i + W_e \text{Onehot}(i)$$

$$i \in [0, 1, 2, \ldots, L - 1]$$

where the column vector $f_i \in \mathbb{R}^{d_h}$ is the i-th element in the feature sequence f. $W_e \in \mathbb{R}^{d_h \times L}$ is a trainable position-embedding weight matrix. The one-hot function converts the position of f_i to the vector in which the i-th element is one and others are zeros. We feed the weighted sum to the Multi-head attention layer as the input tokens.

The self-attention uses dot-product of the d_h-dimensional query and key to relate the features in different positions. The Transformer computes the attention function on a set of queries, keys, and values simultaneously with the matrices Q, K, and V. We set the result of matrix multiplication QK^T as R. The element $r_{i,j}$ in R is the dot product of the i th row in Q and the j-th column in K, measuring the correlation between the j-th feature and the i-th feature. We apply a character-level mask to the matric R. The element of the R filtered by the character-level mask is:

$$CLmask(r_{i,j}) = \begin{cases} -10^{10} & \|i - j\| > T_e \\ r_{i,j} & \|i - j\| \le T_e \end{cases}$$

where T_e is the character-level mask range in the encoder. The f_i only attends to $\left[f_{i-T_e}, f_{i-T_e+1}, f_{i-T_e+2}, \ldots, f_{i-1}, f_i, f_{i+1}, \ldots, f_{i+T_e}\right]$ due to the semantic absence in the sequence,. As the feature sequence f is subsampled from the CAPTCHA image, the upper bound of T_e / L is the W_C / W, W_C is the width of the widest character. The self-attention operation with character-level mask is calculated as:

$$Attention(Q, K, V) = softmax\left(\frac{CLmask(R)}{\sqrt{D}}\right)V$$

The i th feature f_i^o of the output sequence is denoted as:

$$f_i^o = \sum_{j=i-T_e}^{i+T_e} r'_{i,j} v_j$$

where $r'_{i,j}$ is the output of the softmax function and v_j is the j th row of the matrix V. We set the threshold T_e as:

$$T_e = \lceil \frac{1}{N} \times \frac{W_c}{W} \times L \rceil$$

where $N = 2$ is the number of identical layers stacked in the encoder. In addition to attention sub-layers, a fully connected feed-forward network is placed at the top of each layer, which is applied to f_i^o separately.

Decoder. We firstly apply token embedding to the text sequence and obtain the tokens sequence of size $[N_c, h_d]$. Then we introduce position information of the sequence:

$$t_i = \sqrt{d_h} \bullet W_t Onehot(t_i) + W_d Onehot(i)$$

$$i \in [0, 1, 2, \ldots, N_c - 1]$$

where t_i is the index of the i-th character in the text sequence, $W_t \in \mathbb{R}^{d_h \times C}$ and $W_d \in \mathbb{R}^{d_h \times N_c}$ are trainable weight matrices. We similarly apply character-level mask to the self-attention sub-layer in the decoder. Note that the character-level range in decoder $T_d = 1$ and the mask should prevent positions from attending to subsequent positions simultaneously. As a result, the character-level mask forces t_i to attend to t_{i-1} and t_i only. The i th feature t_i^o of the output sequence can be simply denoted as:

$$t_i^o = r'_{i,i-1} v_{i-1} + r'_{i,i} v_i$$

Different from the original Transformer, the self-attention module in the decoder is a single layer to learn the dependency on the previous one text token, not involved in the stack of identical layers.

The structure placed on the top of the self-attention module contains a stack of N identical layers. As shown in Fig. 2, the first layer is an encoder-decoder attention layer, the queries come from the previous decoder layer, and the keys and values come from the output of the encoder. We use no mask in this attention module to allow t_i to attend to all positions in the input sequence.

4   Experiment

4.1   Performance on Different Training Set Sizes

To evaluate the generalization ability of our model, we use different sizes of the training set to train the model and validate the performance on the same test set. In this experiment, we collect 55,000 Sina CAPTCHAs as the dataset. Sina CAPTCHA is the representative text-based CAPTCHA with mixed security features shown in Table 1. The alphanumeric set of Sina CAPTCHA contains 24 categories, and large degrees of rotation and distortion are applied. As shown in table, the neighboring characters in Sina CAPTCHA share the overlapped strokes and the confusing pairs are difficult to recognize.

We label the CAPTCHAs manually and use 5,000 of them for validation. This experiment uses character accuracy to calculate the ratio of correctly recognized characters to all characters and sequence accuracy to measure the success rate of breaking the CAPTCHA with each character correctly recognized simultaneously. Table 4 shows the experimental results.

With only 600 training samples, our model achieves character accuracy of 92.04% and sequence accuracy of 73.22%. As the size of training set increases to 1,000, 6,000, 10,000, the sequence accuracy increases to 85.21%, 93.27%, 93.84% respectively. This result shows that our model can break Sina CAPTCHA in one or two tries by simply collecting 600 samples. The generalization ability on a small training set is superior to prior end-to-end networks.

Table 4 Character accuracy and sequence accuracy achieved on different sizes of Sina CAPTCHA training set

Training set size	Character accuracy	Sequence accuracy
50,000	0.9851	0.9478
20,000	0.9822	0.9365
10,000	0.9829	0.9384
8,000	0.9815	0.9351
6,000	0.9810	0.9327
4,000	0.9791	0.9262
2,000	0.9669	0.8845
1,000	0.9570	0.8521
800	0.9465	0.8152
600	0.9204	0.7322
400	0.6974	0.2739
200	0.1558	0.0004

4.2 The Effect of Mask Range

In this section, we evaluate the trained model's performance using different mask range T_e to explore the reasons for the high accuracy achieved on small training set. We use 600 samples as the training set and 400 samples as the test set. Figure 3 shows the test accuracy achieved on this small dataset.

Our transformer-based model achieves the best test accuracy when we set $T_e = L/4$, consistent with the analysis in Sect. 3.3.1. Note that no character-level mask is applied to the encoder when $T_e = L$. The experiment result proves that the appropriate character-level mask range is crucial to learning the CAPTCHA images' effective representation to improve generalization ability. This is most likely because the character-level mask in the self-attention mechanism forces the encoder to attend to the features extracted from the neighboring characters, ignoring the disturbance from the global features without semantic relevance.

4.3 Larger Scale Attack on CAPTCHAs

To evaluate the robustness of our model to different text-based CAPTCHA schemes, we collect four CAPTCHAs deployed by eBay, Sohu, Netease, and Baidu, respectively. As shown in Table 1, these CAPTCHA schemes cover the widely used security features and also contain confusing pairs. For each scheme, we collect 8,500 CAPTCHA images and manually label them. Similarly, we set the training set size as 600, 1,000, 2,000, 4,000, 8,000 respectively and use 500 samples for test.

Figure 4 shows the character and sequence accuracy on the test set. Our model achieves a high sequence accuracy above 60% trained with only 600 samples except

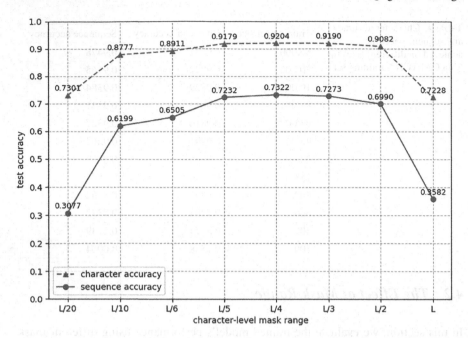

Fig. 3 Test accuracy achieved on the small dataset with different character-level mask ranges

for the Baidu CAPTCHA. However, when the training set slightly increases to 1,000 samples, the sequence accuracy increases rapidly to 77.3%. This is probably because our model requires a bit more samples to learn the representation of the Baidu CAPTCHA with sophisticated fonts and seriously overlapped strokes. The overall performance shows the strong robustness of our model to CAPTCHA schemes with different security features.

4.4 Comparison with Prior Attacks

We further compare our attack against prior attacks on four CAPTCHA. To provide a fair comparison, we use the same dataset or the same CAPTHCA scheme where the prior work was tested on. The result is shown in Table 5, note that the training set size is listed if the attack is based on an end-to-end network.

In this experiment, the performance of our model outperforms the prior work on Sina, eBay, and Baidu CAPTCHAs by achieving a significantly higher success rate. A comparable end-to-end model is the generative adversarial network based approach proposed by Ye et al. [2]. They trained a CAPTCHA synthesizer to automatically generate synthetic CAPTCHAs to learn a base solver. The performance of their model depends on whether the synthesizer can generate high-quality synthetic data.

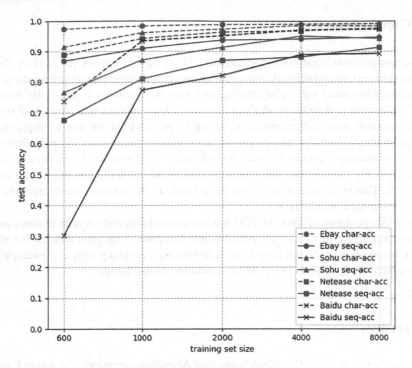

Fig. 4 Performance on eBay, Sohu, Netease, and Baidu dataset

Table 5 Comparing our attack against prior attacks on CAPTCHA schemes where the prior methods were tested on

	Training set size	Sina	eBay	Sohu	Baidu
Zi [9]	2,000	42.0%	/	/	/
Tang [4]	/	51.2%	/	/	/
Gao [19]	/	/	58.8%	/	/
Bursztein [23]	/	/	51.2%	/	/
Liu [24]	10,000	/	/	67.7%	/
Ye [2]	500	52.6%	86.6%	92%	/
Qing [25]	12,000	/	/	/	43.1%
Our model	1,000	85.2%	90.9%	87.1%	77.3%

Our model achieves a lower accuracy on Sohu CAPTCHA but boosts the success rate from 52.6% to 85.2% on Sina CAPTCHA.

5 Conclusion

This paper presents a transformer-based model to solve text-based CAPTCHAs. Our model is an end-to-end network that simultaneously localizes and recognizes the text sequence. Compared with prior attacks, our approach requires significantly fewer labeled samples to train the model and achieves a high success rate. The novel self-attention mechanism with character-level masks proves to be the key to improving the generalization ability. In our model, the encoder focuses on the neighboring characters sharing overlapped strokes we refer to as confusing pairs, and the decoder attends to only the previous one character due to the semantic absence in the text sequence. This mechanism enables our model to learn the effective representation of the text sequence in CAPTCHAs.

We evaluate the model on CAPTCHA schemes with different security features and achieve a comparable or superior performance compared with prior attacks. For the attack based on our model is robust and achieves high accuracy with a few samples, the security of text-based CAPTCHAs should be reconsidered.

References

1. von Ahn L et al (2003) CAPTCHA: using hard AI problems for security. In: Biham E (ed) Advances in cryptology—EUROCRYPT 2003. Springer Berlin Heidelberg, Berlin, Heidelberg, pp 294–311
2. Ye G et al (2018) Yet another text captcha solver: a generative adversarial network based approach. In: Proceedings of the 2018 ACM SIGSAC conference on computer and communications security. CCS '18. Association for Computing Machinery, Toronto, Canada, pp 332–348
3. Starostenko O et al (2015) Breaking text-based CAPTCHAs with variable word and character orientation. Pattern Recogn 48(4):1101–1112
4. Tang M et al (2018) Research on deep learning techniques in breaking text-based captchas and designing image-based captcha. IEEE Trans Inform Forensics Secur 13(10):2522–2537
5. Gao H et al (2017) Research on the security of Microsoft's two-layer captcha. IEEE Trans Inform Forensics Secur 12(7):1671–1685
6. Chellapilla K et al (2005) Computers beat humans at single character recognition in reading based human interaction proofs (HIPs). In: CEAS 2005—second conference on email and anti-spam. Stanford University, California, USA
7. Rui C et al (2013) A novel LSTM-RNN decoding algorithm in CAPTCHA recognition. In: 2013 third international conference on instrumentation, measurement, computer, communication and control, pp 766–771
8. Shu Y, Xu Y (2019) End-to-end captcha recognition using deep CNN-RNN network. In: 2019 IEEE 3rd advanced information management, communicates, electronic and automation control conference (IMCEC), pp 54–58
9. Zi Y et al (2019) An end-to-end attack on text captchas. IEEE Trans Inf Forensics Secur 15:753–766
10. Vaswani A et al (2017) Attention is all you need. In: Advances in neural information processing systems, pp 5998–6008
11. Ding L, Wu D, Tao D (2021) Improving neural machine translation by bidirectional training. In: Proceedings of the 2021 conference on empirical methods in natural language processing, pp 3278–3284

12. Wolf T et al (2020) Transformers: state-of-the-art natural language processing. In: Proceedings of the 2020 conference on empirical methods in natural language processing: system demonstrations, pp 38–45
13. Xu Y et al (2021) ViTAE: vision transformer advanced by exploring intrinsic inductive bias. In: Beygelzimer A et al (eds) Advances in neural information processing systems
14. Dosovitskiy A et al (2021) An image is worth 16×16 words: transformers for image recognition at scale. In: International conference on learning representations
15. Tian Z et al (2016) Detecting text in natural image with connectionist text proposal network. In: European conference on computer vision. Springer, pp 56–72
16. Shi B, Bai X, Yao C (2016) An end-to-end trainable neural network for image-based sequence recognition and its application to scene text recognition. IEEE Trans Pattern Anal Mach Intell 39(11):2298–2304
17. Wojna Z et al (2017) Attention-based extraction of structured information from street view imagery. In: 2017 14th IAPR international conference on document analysis and recognition (ICDAR), vol 1. IEEE, pp 844–850
18. Yan J, El Ahmad AS (2008) A low-cost attack on a Microsoft captcha. In: Proceedings of the 15th ACM conference on computer and communications security. CCS '08. Association for Computing Machinery, Alexandria, Virginia, USA, pp 543–554
19. Gao H et al (2016) A simple generic attack on text captchas. In: The network and distributed system security symposium (NDSS), pp 1–26
20. Goodfellow IJ et al (2014) Multi-digit number recognition from street view imagery using deep convolutional neural networks. arXiv: 1312.6082 [cs.CV]
21. Szegedy C et al (2016) Rethinking the inception architecture for computer vision. In: Proceedings of the IEEE conference on computer vision and pattern recognition, pp 2818–2826
22. UmaMaheswari P et al (2020) Designing a text-based CAPTCHA breaker and solver by using deep learning techniques. In: 2020 IEEE international conference on advances and developments in electrical and electronics engineering (ICADEE), pp 1–6
23. Bursztein E et al (2014) The end is nigh: generic solving of text-based CAPTCHAs. In: Proceedings of the 8th USENIX conference on offensive technologies. WOOT'14. USENIX Association, San Diego, CA, p 3
24. Liu K, Zhang R, Qing K (2017) CNN for breaking text-based CAPTCHA with noise. In: Falco CM, Jiang X (eds) Ninth international conference on digital image processing (ICDIP 2017), vol 10420. International society for optics and photonics. SPIE, pp 615–619
25. Qing K, Zhang R (2017) A multi-label neural network approach to solving connected CAPTCHAs. In: 2017 14th IAPR international conference on document analysis and recognition (ICDAR), vol 01, pp 1313–1317

A New Approach to Image Classification Dataset Privacy-Preserving with Deep Neural Network

Toan Pham Van, Thanh Nguyen Tung, Linh Doan Bao, Duc Tran Trung, Quang Nguyen Hung, and Thanh Ta Minh

Abstract Deep learning technology has made great achievements in some fields such as computer vision, natural language understanding, speech processing, etc. However, a big challenge when using deep learning models is the require a large amount of data. Meanwhile, most of the data belong to organizations or personalities that cannot be public due to privacy. In this paper, we propose a new approach for data publishers to conceal the original dataset but still ensure the features for training deep learning models. Our framework has three advantages. First, we proposed a neural network as a encoder to hide a full-size color image within a noisy image of the same size. The noisy images make it impossible to guessable the contents of the original dataset inside. Second, that method is privacy-preserving as the encode method can't be inverted without having an original dataset. Third, we demonstrate that our framework is accuracy-preserving because the encoded image still keeps features for the image classification task. All of our experiments based on a trade-off between two factors: the number of features retained (information loss), and the difficulty to invert the original dataset (privacy loss). Our code and pre-trained models are available at: https://github.com/sun-asterisk-research/privacy-dataset-research.

Keywords Dataset privacy-preserving · Image classification · Deep neural network

1 Introduction

1.1 Overview

In recent years, data-related technologies such as data mining, data analysis, artificial intelligence have grown rapidly. Accordingly, the data security problems are gradually more concerned [1, 2]. In machine learning, the quality of samples in the

T. P. Van · T. N. Tung · L. D. Bao (✉) · D. T. Trung · Q. N. Hung
Sun Asterisk R&D Lab, Hanoi, Vietnam
e-mail: doan.bao.linh@sun-asterisk.com

T. T. Minh
Le Quy Don Technical University, 236 Hoang Quoc Viet, Bac Tu Liem, Ha Noi, Vietnam

dataset directly affects the performance of the algorithm. However, in fact, using a dataset always requires data privacy constraints. The concern about the leakage of sensitive data has become a barrier for data owners when they intend to publicize their dataset. A dataset privacy-preserving pipeline can be simply defined as follows: Before the data owner publishes their data, they encode the original data in a certain way to hide the sensitive information while preserving the features that are critical for building the classification models. Several methods have been proposed such as data randomization approach [3], condensation approach [4]. The effectiveness of this system is usually evaluated with two metrics: loss of privacy and loss of information (the accuracy of the classification model in our experiment). An ideal system will ensure that the values of both loss functions are minimized. However, the two metrics are not well-balanced in many existing perturbation techniques [3, 5]. Privacy loss can be described as the difficulty level in estimating the original value from the encoded data. The information loss typically refers to the amount of importance features preserved about the dataset after the encoded. Depending on the tasks that the model needs to be solved, the information loss can be defined differently. For example, the classification task will need different features than segmentation or object detection task. Especially with the Convolutional Neural Network (CNN), the ability to extract abstract features makes the model learnable even with indistinguishable images by human. In our experiment, the encoded dataset to be completely black images but still gives up to 80% classification accuracy on CIFAR-10 [6] with familiar architecture as ResNet-50.

1.2 Our Contributions

Our main contributions are:

We propose a solution to hide the original data—an important problem in deep learning problems. In this way, the data used in the model will be safe and secure. Data security is extremely important and urgent, especially problems related to personal data.

We have demonstrated that our model generates a privacy image that is very similar to a container image but retains the original image's features for classification purposes. Comparing the original image input and the privacy image input with the same classification model, the difference is not too large, showing that our model, although hiding data, still stores the hidden features of data.

In the following sections, we briefly review some related works involve with our methods. Section 3 illustrates proposed methodology in detail. Our experiments are described in Sect. 4, including datasets description, data preprocessing methods, and model configurations. Section 5 indicates all experiments and following results. Finally, Sect. 6 is the conclusion for our proposed framework.

2 Related Works

2.1 Deep Neural Network

Deep neural network [7] (DNN) is a complex neural structure system consisting of many neural network units in which, in addition to input and output layers, there is more than one hidden layer. Each of these classes will perform a separate type of classification and sorting in a process we call 'feature hierarchy' and each layer takes on a separate responsibility, the output of this class will be the input of the next class.

DNN is built with the purpose of simulating complex human brain activity and is applied in many different fields, bringing success and amazing results to humans. DNN is applied to process and solve real-life problems such as machine translation [8, 9], medical image analysis [10], virtual assistants [11].

2.2 U-Net

Medical image segmentation initiates the use of symmetric architecture **U-net** [12–15]. It is a specific symmetric instance of the encoder-decoder network structure, with skip connections from layers in the encoder to the corresponding layers in the decoder. The encoder decoder networks have been applied to many computer vision tasks, including object detection and semantic segmentation. Typical networks commence with encoder module that compresses feature maps presentation to capture higher semantic information. A decoder module that recovers previous spatial information. For each layer in UNet's decoder, features can be acquired from higher layers or from skip connected layers.

3 Proposed Methods

In this section, we first present our framework for generating privacy-preserving dataset. Then components in our framework will be demonstrated, including Generator, Container, Loss Function.

3.1 Overview

For privacy-preserving, we train the generator to generate the privacy images that look visually same as container images while still carry information needed for classification tasks. Experimentally, we propose the framework for generate privacy

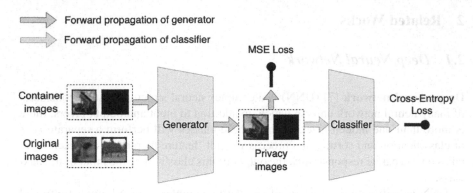

Fig. 1 Overview of our proposed privacy-preserving framework

images and use privacy images for classification task. The overview of our framework
is illustrated in Fig. 1.

3.2 Generator

With the goal of data obfuscation, the Generator model plays an important role in
our framework. Generator takes original images and container images as input, and
outputs images with the desire to be similar to container images, which are privacy
images. With the task of generating privacy images, we experimented with using the
Unet network.

U-Net was created by Olaf Ronneberger, Philipp Fischer, Thomas Brox in 2015
[12]. The main idea is to supplement a regular contract network with successive
layers, where pooling operations are replaced by upsampling operators. These layers
therefore increase the resolution of the output. Furthermore, a successive convolution
layer can then learn to assemble an exact output based on this information.

Together with the Unet network, we minimize the Mean Squared Error loss function so that the privacy images have a high similarity with the Container images—this
proves how good information hiding is. When we tested the generator and classifier
models separately, it was the generator's job to create a privacy image that resembled
the container image as much as possible, regardless of the performance of the classifier model. Also when we test the end-to-end training of the generator and classifier
models, the weight of information hiding will be partially reduced.

3.3 Container

The container is the images used as an overlay to hide the original images, with the expectation that it can completely or almost completely cover the original images for information security. We test that the container is a random images in the dataset, but it may happen that the selected container images and the original images have the same label, which will make the classification model not reliable enough. So, along with the above test, we test container images as black images for evaluation.

3.4 Classifier

We test whether privacy images still store information using a classification model with privacy images as input. We use ResNet-18 as the base network, train and test on it.

ResNet (Residual Network) [16] was introduced to the public in 2015 and even won 1st place in ILSVRC 2015 competition with only 3.57% top 5 error rate, 1st place in ILSVRC and COCO competitions 2015 with ImageNet Detection, ImageNet localization, Coco detection and Coco segmentation. Currently, there are many variants of ResNet architecture with different number of layers such as ResNet-18, ResNet-34, ResNet-50, ResNet-101, ResNet- 152,...With the name ResNet followed by a number indicating the ResNet architecture with a certain number of layers. ResNet-18 is a convolutional neural network that is 18 layers deep. The architecture of ResNet-18 is described by the Table 1.

Table 1 Visualizing ResNet-18 activations

Layer name	Output size	ResNet-18
conv1	$112 \times 112 \times 64$	7×7, 64, stride 2
conv2 $\times$	$56 \times 56 \times 64$	3×3 max pool, stride 2 $(3 \times 3, 64 + 3 \times 3, 64) \times 2$
conv3 x	$28 \times 28 \times 128$	$(3 \times 3, 128 + 3 \times 3, 128) \times 2$
conv4 $\times$	$14 \times 14 \times 256$	$(3 \times 3, 256 + 3 \times 3, 256) \times 2$
conv5 $\times$	$7 \times 7 \times 512$	$(3 \times 3, 512 + 3 \times 3, 512) \times 2$
Average pool	$1 \times 1 \times 512$	7×7 average pool
Fully connected	1000	512×1000 fully connected
softmax	1000	

We train and test on 2 separate sets, one is the original dataset, the other is the privacy dataset. Our team then compare the output of results to see if the difference is significant.

3.5 Loss Function

For Generator, to compare privacy image with original image and container image, we use Mean Squared Error (MSE)loss function to calculate similarity between container image and privacy image. The goal is that the privacy image must really hide the original image. In other words, the privacy image should be similar the container image.

Mean squared error (MSE) is the most used loss function for regression. The loss is calculate by taking mean of overseen data of the squared differences between true and predicted values or writing it as a formula.

$$L_{mse}(y, \hat{y}) = \frac{1}{N} \sum_{i=0}^{N} (y - \hat{y}_i)^2 \tag{1}$$

where $\hat{y}$ is the predicted value.

For Classifier, the task of the model is to classify images with input as privacy images to verify hidden information of privacy images. Cross Entropy (CE) loss function is used for the classifier model.

CE is a loss function that is used in multi-class classification tasks. CE loss, or log loss, measures the performance of a classification model whose output is a probability value between 0 and 1. CE loss increases as the predicted probability diverges from the actual label. CE loss function is calculated as follows

$$L_{ce}(y, \hat{y}) = \frac{1}{N} \sum_{i=0}^{N} y_i \cdot \log \hat{y}_i \tag{2}$$

where $\hat{y}_i$ is the ith scalar value in the model output, y_i is the corresponding target value, and N is the number of scalar values in the model output.

4 Experiments

4.1 Experimental Setup

In all experiments, the hyper-parameters used for training generator are set as follows: the batch size is 128, Adam optimizer is adopted with learning rate 0.1; coefficient used for computing running averages of gradient is 0.5.

In our experiment, CIFA-10 are chosen to be testing dataset. The training batches contain the remaining images in random order, but some training batches may contain more images from one class than another. Between them, the training batches contain exactly 5000 images from each class.

To evaluate whether our decoder model retains image features, we additionally use the CIFAR-100 [6] dataset for training and evaluating the classifier model with the input of the privacy image by our decoder model. Dataset is split with ratio 5:1 (images per class) between training and testing.

4.2 Evaluation Metrics

We evaluate our system in three criterions: the ability of obfuscating original images, the potential of transferring to downstream tasks (i.e. image classification), and robustness to model inversion attacks.

We use pixel-to-pixel mean square error between original and encoded images to measure the difference. The higher the error is, the more difficult human vision can recognize sensitive information in processed images.

$$MSE(I_1, I_2) = \frac{1}{mn} \sum_{i=1}^{m} \sum_{j=1}^{n} |I_1(i, j) - I_2(i, j)|^2$$

For image classification task, we compute model accuracy to evaluate the amount of useful retained features. To prove that our methods can defend against image deobfuscation, we calculate mean square error between output images from decoder and original images. Similar to the first criteria, the high error shows that model inversion attacks cannot reconstruct original data as well as obtain any important information.

4.3 Train Generator and Classifier Separately

Achieving high accuracy generator is a difficult task due to previous reported problems: instability and uncertainty [17], mode collapse [18, 19], unsteady gradient [20].

Our method resolves these difficulties with a divide-and-conquer method, continuously training the generator from lower resolution to high, resulting in faster and more stable training. A simpler version of mini-batch and some other training techniques to discourage the generator are used against collapse mode.

4.4 Multi-task Learning

While forcing generator to recreate container image can hide information well, there is a considerable trade-off in accuracy. In this experiment, we add more constraint to generated images. The purpose of this is to force generated images to contain features needed for classification tasks.

Specifically, we combined classifier on top of generator to create an multi-task learning paradigm.

$$L_{overall} = \beta \cdot L_{MSE} + (1 - \beta) L_{BCE} \tag{3}$$

where β is the hyper-parameter to control the trade-off between privacy-preserving and feature-preserving in the overall loss $L_{overall}$. When training on CIFAR10, the multi-task model is consist of same generator and classifier as training them separately. Whereas, when testing generator on CIFAR100, we use only generator for generating data and use different classifier. Figure 2 shows results of our experiment with different βs

As shown in figure, this multi-task paradigm does improve the accuracy of classification task. For checking privacy-preserving property, we use same method in separate training for decoding. The privacy images produced by this method can be decoded easily as shown in Fig. 3. Therefore, we do not adopt this paradigm in our proposed method.

Fig. 2 Influence of multi-task training

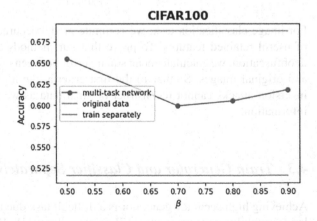

Fig. 3 An attempt to decode
privacy images generated by
multi-task learning paradigm
with β = 0.9. From left to
right: original images,
privacy images, decoded
images

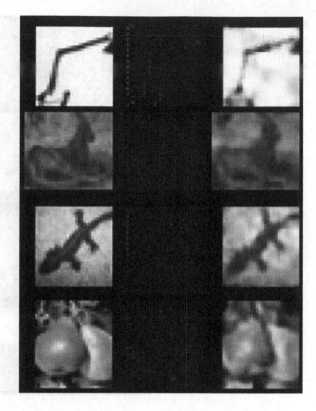

4.5 Influence of Container Image

In our proposed method, we train the generator with different containers. So that,
the generator can work on an arbitrary container image. In this experiment, we use
different strategy for choosing container. We train the generator on CIFAR10. The
privacy dataset is generated from CIFAR100 using trained generator. The purpose
of this is to find the effect of container on classifier and decoder. The black image is
used as base container for comparison.

In the first experiment, we use completely random images as container for generating privacy images as illustrated in Fig. 4. We train the classifier on this privacy
dataset and achieve the accuracy on a par with the proposed method. So using
random images instead of black image as containers does not hurt the accuracy
on classification task.

In the second experiment, we train generator follow multi-task paradigm (β = 0.5)
on CIFAR10. We also use random images as container for generating privacy images.
This still does hurt the accuracy achieve by classifier. We did attempt to decode the
privacy images. We found that the behavior of decoder is almost the same. So using
random images does not improve privacy-preserving property.

Fig. 4 From left to right:
container (random image),
original image, privacy
image

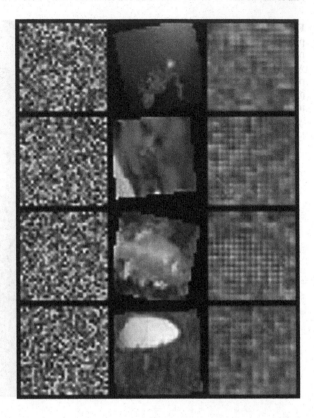

Since privacy-preserving property not improved by using different images as
container, for simplicity we use black pixel image as container in our method.

5 Results

5.1 Generator Model

With the use of Mean Square Error (MSE) loss function to evaluate the similarity
between the container images and the output images of our generator model (privacy
images), the results after using the CIFAR10 dataset to train the generator model with
100 epochs, learning rate 0.01 is illustrated in Table 2. The relatively small MSE loss
means that the container images and the privacy images are quite similar. Figure 5
illustrates a set of 3 images: container image—real image—privacy image—which
is the test result of any 1 image on the test set after training our generator model.

Table 2 Training loss and testing loss of generator model with CIFAR-10 dataset

Dataset	Training loss	Testing loss
CIFAR-10	0.0003204	0.0005509

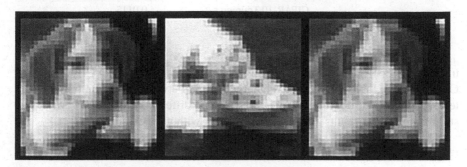

Fig. 5 Triple images: container image—real image—privacy image

Table 3 Result of classifier model with CIFAR-10 and CIFAR-100 dataset

Dataset	Training loss	Training accuracy	Testing loss	Testing accuracy
CIFAR-10	0.02662	0.99106	0.53078	0.9161
Generator CIFAR-10	0.07983	0.97254	1.29643	0.792
CIFAR-100	0.13459	0.95686	2.78468	0.6234
Generator CIFAR-100	0.30858	0.90124	3.65142	0.5076

5.2   Classifier Model

After encoder image to privacy dataset, we train and test another model with privacy dataset as input to evaluate whether our privacy model retains the original image's features to serve the following tasks. Here we test is the classifier task. We use the RestNet18 model as the base classifier model to compare the input as the original image and the input as the privacy image to consider the accuracy and error of the model. Two datasets are used for this classification task, CIFAR-10 (100 epochs) and CIFAR-100 (50 epochs), in which CIFAR-10 was used for Generator task. The results obtained in the Table 3, show that the difference is not too large, the classification model still gives good results. This proves that our Generator model still retains the features of the original image, although not the entire image.

5.3   Decoder Model

Given a trained generator on CIFAR-10, a decoder is then trained on CIFAR-100, since we assume the attacker can not access the true dataset in which generator was

Table 4 Mean square error of decoder model on CIFAR-10 and CIFAR-100 datasets

Dataset	Mean square error
CIFAR-100 Training set	0.0124
CIFAR-100 Test set	0.0138
CIFAR-10 Training set	0.0113
CIFAR-10 Test set	0.0124

trained. We report mean square error on both training and test sets of CIFAR-10 and CIFAR-100 in Table 4. The relatively large margin between results on training and test sets implies the decoder model is overfitted on different datasets, however it can still obtain similar results on the original dataset without accessibility.

6 Conclusions

We have presented a solution for data security for machine learning problems. In this section, we briefly discuss a few observations found in this study and present ideas for future work. By testing image masking with a Unet network with the constraint of minimizing MSE, we have proven that it is possible to hide images while retaining image features. At the same time, it is relatively difficult to decode again from the privacy image to the original image. We came to the conclusion that hiding data using containers is possible, and compared to the original data, the privacy dataset still has value for future tasks.

There are still many problems to be solved, such as scaling between information hiding and original image information storage. Remember that the more information the original image is stored, the less effective the masking will be, and the easier it will be to decode the original image again.

Acknowledgements This work is partially supported by Sun-Asterisk Inc. We would like to thank our colleagues at Sun-Asterisk Inc for their advice and expertise. Without their support, this experiment would not have been accomplished.

References

1. Hanukoglu M, Goldberg N, Rovshitz A, Azaria A (2019) Learning to conceal: a deep learning based method for preserving privacy and avoiding prejudice
2. Zhu J, Kaplan R, Johnson J, Fei-Fei L (2018) Hidden: hiding data with deep networks
3. Agrawal R, Srikant R (2000) Privacy-preserving data mining. In: Proceedings of the 2000 ACM SIGMOD international conference on management of data, pp 439–450
4. Aggarwal CC, Philip SY (2004) A condensation approach to privacy preserving data mining. In: International conference on extending database technology. Springer, pp 183–199

5. Evfimievski A, Gehrke J, Srikant R (2003) Limitingprivacy breaches in privacy preserving data mining. In: Proceedings of the twentysecond ACM SIGMOD-SIGACT-SIGART symposium on Principles of database systems, pp 211–222
6. Krizhevsky A (2009) Learning multiple layers of features from tiny images. Technicalreport
7. Hinton GE, Salakhutdinov RR (2006) Reducing the dimensionality of data withneural networks. Science (New York, N.Y.) 313:504–507
8. Luong MT, Pham H, Manning CD (2015) Effective approaches to attention-based neural machine translation
9. Wu Y, Schuster M, Chen Z, Le QV, Norouzi M, Macherey W, Krikun M, Cao Y, Gao Q, Macherey K, Klingner J, Shah A, Johnson M, Liu X, Kaiser L, Gouws S, Kato Y, Kudo T, Kazawa H, Stevens K, Kurian G, Patil N, Wang W, Young C, Smith J, Riesa J, Rudnick A, Vinyals O, Corrado G, Hughes M, Dean J (2016) Google's neural machine translation system: bridging the gap between human and machine translation
10. Gusarev M, Kuleev R, Khan A, Rivera AR, Khattak AM (2017) Deep learning models for bone suppression in chest radiographs. In: 2017 IEEE conference on computational intelligence in bioinformatics and computational biology (CIBCB), pp 1–7
11. Campagna G, Ramesh R (2017) Deep almond: a deep learning-basedvirtual assistant [language-to-code synthesis of trigger-action programs using seq 2 seq neural networks]
12. Ronneberger O, Fischer P, Brox T (2015) U-net: convolutional networks for biomedical image segmentation
13. Alom MZ, Hasan M, Yakopcic C, Taha TM, Asari VK (2018) Recurrent residual convolutional neural network based on u-net (r2u-net) for medical image segmentation
14. Oktay O, Schlemper J, Folgoc LL, Lee M, Heinrich M, Misawa K, Mori K, McDonagh S, Hammerla NY, Kainz B, Glocker B, Rueckert D (2018) Attention u-net: learning where to look for the pancreas
15. Zhou Z, Siddiquee MR, Tajbakhsh N, Liang J (2018) Unet++: ass nested u-net architecture for medical image segmentation
16. He K, Zhang X, Ren S, Sun J (2015) Deep residual learning for image recognition
17. Salimans T, Goodfellow I, Zaremba W, Cheung V, Radford A, Chen X (2016) Improved techniques for training gans
18. Berthelot D, Schumm T, Metz L (2017) Began: boundary equilibrium generative adversarial networks
19. Metz L, Poole B, Pfau D, Sohl-Dickstein J (2016) Unrolled generative adversarial networks
20. Arjovsky M, Bottou L (2017) Towards principled methods for training generative adversarial networks

Differential Synchronization Sequence Design for OFDM Systems with Large Doppler Shift

Feng Zhu, Jie Ni, Xulin Shi, Xiqi Gao, and Conghao Zhang

Abstract The Orthogonal Frequency Division Multiplexing (OFDM) is widely used in many systems due to its high spectrum utilization. However, the orthogonality of subcarriers in OFDM can be easily destroyed by the inter-carrier interference caused by Doppler frequency offset. In the synchronization of OFDM systems with large Doppler shift, differential cross-correlation is a common method to reduce the influence of frequency offset, but it is sensitive to noise. This paper proposes a differential Zadoff-Chu (DZC) sequence with constant amplitude and good correlation. Most importantly, the sequence obtained by the conjugate multiplication of adjacent symbols of the proposed sequence is a ZC sequence, which has constant amplitude and zero autocorrelation. Simulation results show that the proposed DZC sequence has a significant performance improvement compared with the traditional ZC sequence in the synchronization method based on differential cross-correlation.

Keywords Differential cross-correlation · DZC sequence · OFDM · Synchronization

1 Introduction

Orthogonal Frequency Division Multiplexing (OFDM) is widely used in LTE, 5G, WIFI, DVB-T, and other high throughput transmission systems due to its high spectrum utilization [1–5]. However, OFDM systems are very sensitive to timing deviation and carrier frequency offset (CFO), especially in some scenarios with high-dynamic and large Doppler frequency shift, such as low earth orbit (LEO) satellite, high-speed aircraft, high-speed railway scenes, etc. [6–8]. Timing deviation will lead to inter-symbol interference (ISI), and CFO will cause inter-carrier interference

F. Zhu (✉) · J. Ni · X. Shi · X. Gao
National Mobile Communications Research Laboratory, Southeast University, Nanjing 210096, Jiangsu, China
e-mail: fzhu@seu.edu.cn

C. Zhang
CAS Key Laboratory of Wireless-Optical Communications, USTC, Hefei 230027, Anhui, China

© The Author(s), under exclusive license to Springer Nature Singapore Pte Ltd. 2023   33
Y. Ma (ed.), *Advanced Theory and Applications of Engineering Systems Under the Framework of Industry 4.0*, https://doi.org/10.1007/978-981-19-9825-6_3

(ICI) [9]. Therefore, timing and frequency synchronization is a vital step for OFDM systems.

Many methods have been presented for synchronization in OFDM systems, including the autocorrelation method, cross-correlation method, and the combination of the two methods [10, 11]. Autocorrelation methods usually use the cyclic prefix (CP) of the OFDM symbol to detect timing deviation and frequency offset [12, 13]. Since CP is added in an OFDM symbol, these methods are only suitable for cases where there is only a fractional frequency offset, that is, the CFO is less than the subcarrier interval. Cross-correlation methods use local synchronization sequence to estimate the timing and frequency offset [14, 15]. However, in the scenarios with high-dynamic and large Doppler offset, the position and amplitude of the peak value will change greatly which will lead to the rapid degradation of the performance of these methods. In [16], the symmetric property of Zadoff-Chu (ZC) sequence and the conjugate of ZC sequence was used to improve timing synchronization performance.

To reduce the influence of frequency offset, the researchers proposed a differential cross-correlation method, in which the received signal is multiplied by the conjugate of the adjacent symbol so that the phase accumulation caused by frequency offset can be avoided in the cross-correlation with the local differential sequence [17, 18]. However, the sequence after differential operation destroys the correlation properties of the original sequence, so that the peak value is no longer sharp, and the curve after cross-correlation is almost triangular, which makes the peak detection more susceptible to noise.

In this paper, a special differential Zadoff-Chu (DZC) sequence is designed, and the sequence after the differential operation is a ZC sequence, which can greatly improve the synchronization performance in the differential cross-correlation method. The proposed sequence can be applied to LEO satellite, high-speed aircraft, high-speed railway, and other communication systems with large Doppler frequency shift.

The remainder of this paper is organized as follows. The system model is introduced in Sect. 2. Section 3 presents the proposed method. Simulation results are shown in Sect. 4 and the paper is concluded in Sect. 5.

2 System Model

The OFDM symbol can be expressed as:

$$s(n) = \frac{1}{\sqrt{N}} \sum_{k=0}^{N-1} S(k) e^{j 2\pi kn/N}, n = 0, 1, \ldots, N-1, \tag{1}$$

where N is the total number of subcarriers, $X(k)$ represents the data symbol transmitted on the kth subcarrier. Then, CP is added at the beginning of each OFDM symbol to avoid ICI, and the CP can be obtained by

$$s(n) = s(n + N), n = -N_g, \ldots, -1, \tag{2}$$

where N_g is the length of CP.

Considering the delay and frequency offset, the received signal under the additive white Gaussian noise (AWGN) channel can be expressed as

$$r(n) = s(n - \theta)e^{j2\pi n\varepsilon/N} + w(n), \tag{3}$$

where ε denotes the normalized CFO with respect to the subcarrier interval, θ represents the normalized timing deviation with respect to the sampling interval, and $w(n)$ is the zero-mean complex additive white Gaussian noise.

The constant amplitude zero autocorrelation (CAZAC) sequence which has a constant envelope and good autocorrelation is often used as a synchronization sequence. For example, the ZC sequence is a typical CAZAC sequence with zero autocorrelation and good cross-correlation, which has been applied in LTE, 5G, and other OFDM systems [19, 20]. ZC sequence can be expressed as:

$$z_u(n) = e^{-j\frac{\pi un(n+q)}{N_{ZC}}}, n = 0, 1, \ldots, N_{ZC} - 1, \tag{4}$$

where $q = N_{ZC}$ mod 2, N_{ZC} is the length of ZC sequence, u denotes the root index which is an integer satisfying that u and N_{ZC} are co-prime.

The differential cross-correlation method is commonly used for synchronization in scenarios with large Doppler frequency offset. Multiply the received signal in (3) by the conjugate of its adjacent symbol, thus obtaining

$$
\begin{aligned}
R(n) &= r^*(n)r(n + 1) \\
&= e^{j2\pi\varepsilon/N} s^*(n - \theta)s(n - \theta + 1) + w'(n),
\end{aligned}
\tag{5}
$$

where $w'(n)$ is the equivalent noise after the differential operation. Then, calculate the cross-correlation of $R(n)$ and the local differential sequence $s^*(n)s(n+1)$. According to (5), the frequency offset ε in $R(n)$ is no longer related to index n. Therefore, the frequency offset no longer leads to phase accumulation in the cross-correlation results, but only appears as a fixed phase $e^{j2\pi\varepsilon/N}$.

Although the differential cross-correlation reduces the effect of frequency offset, the sequence after differential operation destroys the correlation characteristics of the original sequence. As shown in Fig. 1, the curve of differential cross-correlation by using ZC sequence is no longer sharp but similar to a triangle, making the peak judgment more sensitive to the noise.

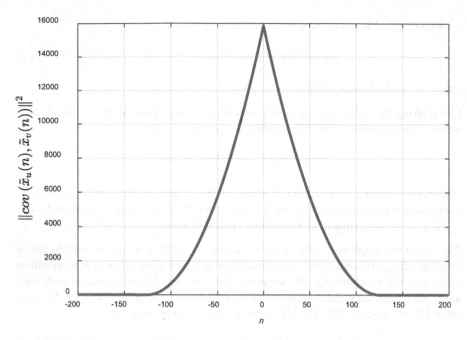

Fig. 1 Differential cross-correlation result of the ZC sequence, $N_{ZC} = 127$

3 Proposed Mothed

3.1 Differential ZC Sequence Design

To improve the differential correlation performance of the traditional ZC sequence, a differential-ZC sequence is specifically constructed, which can keep good auto-correlation and cross-correlation after the differential operation. The proposed DZC sequence $x_u(n)$ needs to satisfy the following requirements:

- The sequence is constant amplitude, i.e., $\|x_u(n)\|^2 = 1, n = 0, 1, \ldots, N_{DZC} - 1$.
- The sequence after the differential operation is a ZC sequence, that is, $\tilde{x}_u(n) = x_u^*(n-1)x_u(n)$ is a ZC sequence of length $N_{DZC} - 1$ and root u, where $n = 1, \ldots, N_{DZC} - 1$.

Based on the above definitions, let

$$\frac{x_u(n)}{x_u(n-1)} = z_u(n-1). \tag{6}$$

Then, $x_u(n)$ can be written as:

$$x_u(n) = z_u(n-1)x_u(n-1) = x_u(0) \prod_{m=0}^{n-1} z_u(m), \tag{7}$$

where the initial value $x_u(0)$ can be defined to be equal to 1. Substituting (4)–(7), $x_u(n)$ can be expressed as:

$$x_u(n) = e^{-j\frac{\pi u(n-1)n(2n-1+3q)}{6(N_{DZC}-1)}}, \tag{8}$$

where $q = (N_{DZC} - 1) \bmod 2$, u is the root index.

3.2 Correlation of Proposed Sequence

According to the definition of the proposed sequence, the sequence $\tilde{x}_u(n)$ after the differential operation is a ZC sequence, which has zero autocorrelation and good cross-correlation as shown in Fig. 2.

In addition to the advantage of differential cross-correlation, the proposed DZC sequence also has good correlation characteristics in the time domain and frequency domain. According to the generation formula (8), the DZC sequence is a constant amplitude sequence, which meets the requirement of low peak-to-average ratio (PAPR) of OFDM systems. As shown in Fig. 3, the constructed DZC sequence also has good autocorrelation and cross-correlation properties.

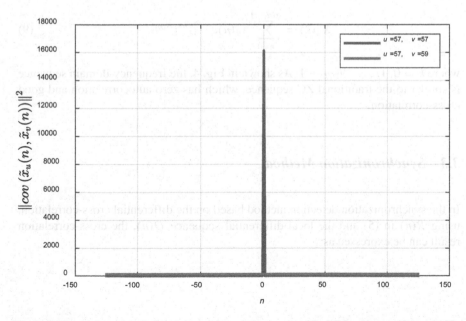

Fig. 2 Differential autocorrelation and cross-correlation results of the DZC sequence, $N_{DZC} = 128$

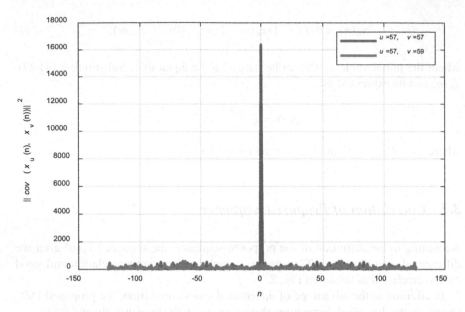

Fig. 3 Autocorrelation and cross-correlation results of the DZC sequence, $N_{\text{DZC}} = 128$

Through discrete Fourier transform (DFT), the frequency-domain expression of the proposed sequence can be written as

$$X_u(k) = \sum_{n=0}^{N_{\text{DZC}}-1} x_u(n)e^{-j2\pi nk/N_{\text{DZC}}}, \tag{9}$$

where $k = 0, 1, \ldots, N_{\text{DZC}} - 1$. As shown in Fig. 4, the frequency-domain sequence is similar to the traditional ZC sequence, which has zero autocorrelation and good cross-correlation.

3.3 Synchronization Method

In the synchronization detection method based on the differential cross-correlation, using $R(n)$ in (5) and the local differential sequence $Q(n)$, the cross-correlation result can be expressed as:

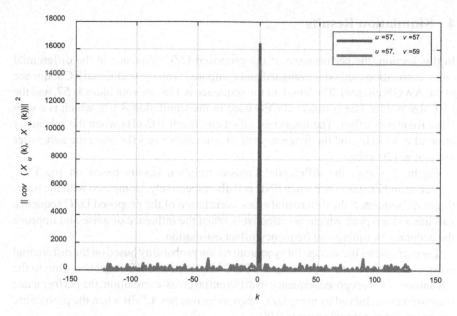

Fig. 4 Autocorrelation and cross-correlation results of the frequency-domain sequence, $N_{\text{DZC}} = 128$

$$J(p) = \sum_{n=0}^{N-2} R(n+p)Q^*(n)$$

$$= e^{j2\pi\varepsilon/N} \sum_{n=0}^{N-2} \left(s^*(n+p-\theta)s(n+p-\theta+1)\right)\left(s^*(n)s(n+1)\right)^* + w''(n),$$

$$(10)$$

where $Q(n) = s^*(n)s(n+1)$. The frequency offset no longer causes the change of the position and amplitude of the peak value, but only affects the phase of the cross-correlation result, which can be used to estimate the CFO. Thus, the estimated timing deviation and frequency offset can be expressed as

$$\hat{\theta} = \arg\max_p |J(p)|^2$$

$$\hat{\varepsilon} = \frac{N}{2\pi} angle\left(J(\hat{\theta})\right).$$

$$(11)$$

4   Simulation Results

In this section, the performance of the proposed DZC sequence in the differential cross-correlation method is evaluated and compared with the traditional ZC sequence in the AWGN channel. The length of the sequence is 128, the root index is 57, and the FFT size is 256. The normalized CFO used in the simulation is 6.8, which is a very large frequency offset. The frequency offset can reach 102 kHz when the subcarrier interval is 15 kHz, and the frequency offset can reach 816 kHz when the subcarrier interval is 120 kHz.

Figure 5 shows the differential cross-correlation results based on the DZC sequence and ZC sequence when SNR is 0 dB respectively. Compared with the traditional ZC sequence, the differential cross-correlation of the proposed DZC sequence still has a sharp peak which can effectively resist the influence of noise and improve the accuracy of timing and frequency offset estimation.

Figure 6 shows the successful synchronization probability based on the differential cross-correlation method by using the DZC sequence and ZC sequence. Due to the advantage of the proposed sequence in differential cross-correlation, the performance improvement achieved by using DZC sequences reaches 4.5 dB when the probability of successful synchronization is 0.99.

Figure 7 compares the mean square error (MSE) of timing estimation based on the differential cross-correlation method. The simulation result shows that the proposed sequence has superior performance to the traditional ZC sequence. Since the peak of differential cross-correlation by using ZC sequence is not sharp, the peak position fluctuates more severely under noise interference.

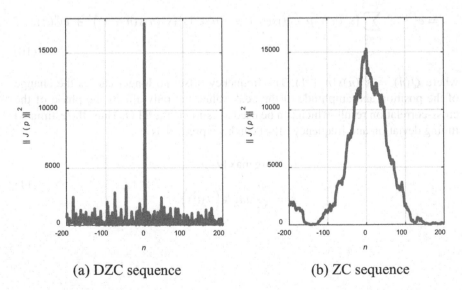

(a) DZC sequence   (b) ZC sequence

Fig. 5 Differential cross-correlation results when SNR is 0 dB. **a** DZC sequence **b** ZC sequence

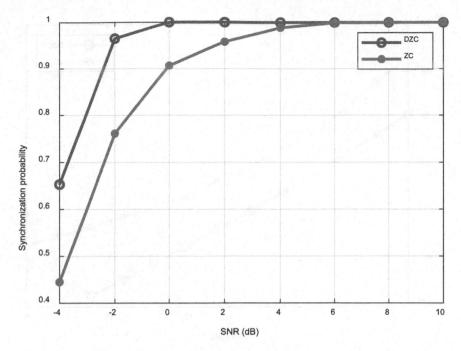

Fig. 6 Successful synchronization probability based on differential cross-correlation method

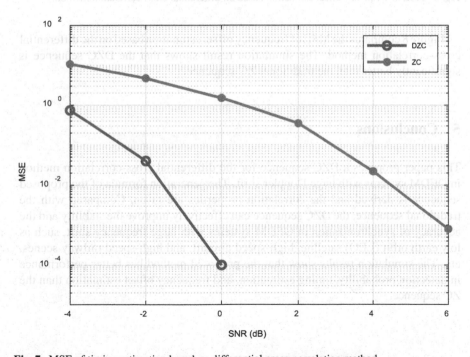

Fig. 7 MSE of timing estimation based on differential cross-correlation method

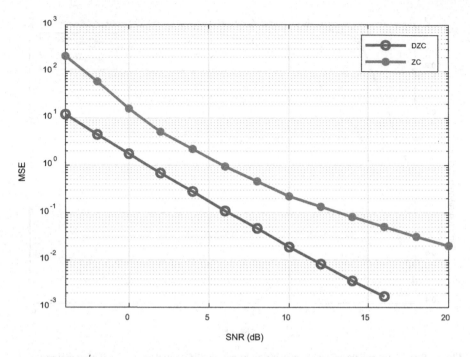

Fig. 8 MSE of frequency offset estimation based on differential cross-correlation method

Figure 8 compares the MSE of frequency offset estimation based on the differential cross-correlation method. The simulation result shows that the DZC sequence is significantly better than the traditional ZC sequence.

5 Conclusions

This paper proposes a DZC sequence for the differential cross-correlation method in OFDM systems with large Doppler shift. The generation formula of the proposed sequence is derived and the correlation properties are given. Compared with the traditional sequence, the DZC sequence can effectively improve the stability and the accuracy of synchronization in OFDM systems with large frequency shift, such as low earth orbit (LEO) satellite, high-speed aircraft, and high-speed railway scenes, etc. The simulation results show that the proposed method has better performance in detection probability, timing estimation, and frequency offset estimation than the ZC sequence.

References

1. Nassralla MH, Mansour MM, Jalloul LMA (2016) A low-complexity detection algorithm for the primary synchronization signal in LTE. IEEE Trans Veh Technol 65(10):8751–8757
2. Wei T, Liu W, Tseng C, Jou S (2009) Low complexity synchronization design of an OFDM receiver for DVB-T/H. IEEE Trans Consum Electron 55(2):408–413
3. Omri A, Shaqfeh M, Ali A, Alnuweiri H (2019) Synchronization procedure in 5G NR systems. IEEE Access 7:41286–41295
4. Zhu F, Ba T, Zhang Y, Gao X, Zhai S, Gong X (2021) Low-complexity detection method based on channel matrix periodic N-diagonal equivalence for uplink MU-MIMO of multi-beam satellite communication systems. Int J Satell Commun Netw 39(5):509–523
5. He Y, Chen Y, Hu Y, Zeng B (2020) WiFi vision: sensing, recognition, and detection with commodity MIMO-OFDM WiFi. IEEE Internet Things J 7(9):8296–8317
6. Wu J, Fan P (2016) A survey on high mobility wireless communications: challenges, opportunities and solutions. IEEE Access 4:450–476
7. Wang W, Tong Y, Li L, Lu A-A, You L, Gao X (2019) Near optimal timing and frequency offset estimation for 5G integrated LEO satellite communication system. IEEE Access 7:113298–113310
8. Stevens BW, Younis MF (2021) Detection algorithm for cellular synchronization signals in airborne applications. IEEE Access 9:55555–55566
9. Ai B, Yang Z-X, Pan C-Y, Ge J-H, Wang Y, Zhen L, (2006) On the synchronization techniques for wireless OFDM systems. IEEE Trans Broadcast 52(2)236–244
10. Zeng R, Huang H, Yang L, Zhang Z (2018) 'Joint estimation of frequency offset and Doppler shift in high mobility environments based on orthogonal angle domain subspace projection. IEEE Trans Veh Technol 67(3):2254–2266
11. Omri A, Shaqfeh M, Ali A, Alnuweiri H (2019) Synchronization procedure in 5G NR systems. IEEE Access 7:41286–41295
12. Sandell M, Beek JVD, Börjesson PO (1995) Timing and frequency synchronization in OFDM systems using the cyclic prefix. In: Proceedings of international symposium synchronization, Essen, Germany, pp 16–19
13. Fusco T, Tanda M (2009) Blind synchronization for OFDM systems in multipath channels. IEEE Trans Wirel Commun 8(3):1340–1348
14. Morelli M, Moretti M (2016) A robust maximum likelihood scheme for PSS detection and integer frequency offset recovery in LTE systems. IEEE Trans Wirel Commun 15(2):1353–1363
15. Nassralla MH, Mansour MM, Jalloul LMA (2016) A low-complexity detection algorithm for the primary synchronization signal in LTE. IEEE Trans Veh Technol 65(10):8751–8757
16. Zhao Y, Cao J, Li Y (2018) An improved timing synchronization method for eliminating large doppler shift in LEO satellite system. In: Proceedings of IEEE 18th international conference communication technology (ICCT), pp 762–766
17. Liu Y, Yu H, Ji F, Chen F, Pan W (2014) Robust timing estimation method for OFDM systems with reduced complexity. IEEE Commun Lett 18(11):1959–1962
18. Zhen L, Bashir AK, Yu K, Al-Otaibi YD, Foh CH, Xiao P (2021) Energy-efficient random access for LEO satellite-assisted 6g internet of remote things. IEEE Internet Things J 8(7):5114–5128
19. Mansour M (2009) Optimized architecture for computing Zadoff-Chu sequences with application to LTE. In: Proceedings of IEEE global telecommunications conference (GLOBECOM). IEEE, pp 1–6
20. Schreiber G, Tavares M (2018) 5G new radio physical random access preamble design. In: Proceedings of IEEE 5G World Forum (5GWF), Silicon Valley, CA, USA, pp 215–220

To the Conditions for Using Magnetic Susceptibility Data of Dispersed Materials to Determine the Susceptibility of Their Particles

Sandulyak Anna, Sandulyak Darya, Ershova Vera, Polismakova Maria, and Sandulyak Alexander

Abstract Attention is drawn to the relevance of the problem not yet solved to date: about acquisition of data on the magnetic susceptibility χ of "isolated" magnetically active particles of small sizes, in particular, iron-containing particles present in loose and liquid media subjected to targeted magnetic action in industrial technologies, biotechnologies and to eliminate environmental problems. The approach proposed for this is based on obtaining data on the magnetic susceptibility of a sample of a dispersed material. The dispersed phase consists of iron-containing particles. The sample should be moderately rarefied. At the same time, the volume fraction of γ particles in the sample should not exceed such a (criteria) value $\gamma = [\gamma]$, when the mutual separation of the particles is achieved, which practically eliminates their magnetic interaction. Then the value of χ can be additionally estimated from the relation. The desired value $[\gamma] = 0.15\ldots0.2$ follows from the examples of informative concentration dependences of magnetic susceptibility for different dispersed samples with a dispersed phase, in particular, particles of natural magnetite, ferroimpurity particles (isolated from sugar and semolina), particles of a carbon sorbent modified (with inclusions of magnetite and maghemite).

Keywords Dispersed material · Magnetoactive particles · Magnetic susceptibility · Field strength

1 Introduction

In the preparation and development of scientific and practical solutions for the target magnetic effect on magnetically active iron-containing particles present in loose

S. Anna (✉) · S. Darya · P. Maria · S. Alexander
MIREA – Russian Technological University, Stromynka Str., 20, 107076 Moscow, Russian Federation
e-mail: a.sandulyak@mirea.ru

E. Vera
Moscow Polytechnic University, B. Semenovskaya Str., 38, 107023 Moscow, Russian Federation

© The Author(s), under exclusive license to Springer Nature Singapore Pte Ltd. 2023   45
Y. Ma (ed.), *Advanced Theory and Applications of Engineering Systems Under the Framework of Industry 4.0*, https://doi.org/10.1007/978-981-19-9825-6_4

and liquid media, the problem of obtaining data on the magnetic susceptibility of these particles is always relevant, and one of the priorities (not yet solved). This is indicated, first of all, by the basic expression for the ponderomotive (magnetic) force [1–7] acting in an inhomogeneity ($gradH \neq 0$) magnetic field on an isolated small particle of volume V_i, which at a field strength H, has a magnetic susceptibility χ:

$$F_i = \mu_0 H\, grad\, H \cdot \chi \cdot V_i, \tag{1}$$

where $\mu_0 = 4\pi \times 10^{-7}$ H/m—magnetic constant.

Below are some examples of targeted magnetic action on magnetically active iron-containing particles present in many media. These examples testify to the scale significance of the problem of obtaining data on the magnetic susceptibility χ of such particles.

Thus, this task is very relevant in the enrichment of iron-bearing ores, when the concentrate of minerals is obtained by dry and/or wet magnetic separation. The composition of such particles is an arbitrary combination of not only magnetically active iron (not atomic) and its magnetic compounds (magnetite Fe_3O_4, maghemite γ-Fe_2O_3, etc.), but also many other chemical elements and compounds. Due to the actually complicated composition of iron-containing particles, the variability of the fractional presence of magnetically active components in them (even for related ores), the task of obtaining data on their magnetic susceptibility for choosing modes and techniques of magnetic separation becomes very important.

A similar problem is just as relevant in the case when the option of using magnetic separation (for fine magnetic separation: magnetic filtration) is considered in order to remove unwanted iron-containing impurities from it. Such impurities (as a rule, the consequences of wear and corrosion of steel equipment) adversely affect the quality of the environment, production technology, reliability, safety and durability of the equipment. This applies to a very wide range of media, in particular, raw materials and ingredients for the production of ceramics, glass products and food products, fuels and lubricants, waters and condensates of thermal and nuclear power plants, ammonia for the production of nitric acid, ammonia water (pure for analysis), circulating water of a rolling mill, etc. [8, 9]. The highly undesirable presence of these impurities in various working environments is evidenced by the numerous regulatory framework of Russia on the corresponding limitation of their presence [9] (usually in terms of iron itself or its oxides). In each specific case (even for productions of the same type), the composition of iron-containing impurities is, as in the previous example, an arbitrary combination of not only iron and its magnetic compounds (magnetite, maghemite, etc.), but also many other chemical elements and connections. It depends on the composition of the working media, the condition of the equipment, production modes, etc. The problem of obtaining data on the magnetic susceptibility of such particles, the composition of which is complicated by impurity particles, becomes really relevant.

To no lesser extent, this problem is relevant for the so-called magnetic (more precisely, magnetically active) carbon sorbents, which are used to solve environmental problems such as water purification from heavy metals, organic compounds, dyes, pharmaceuticals and other pollutants [10]. The basis of such, i.e. modified, carbon sorbents are particles of activated powdered carbon. These particles are additionally "impregnated" with magnetically active iron compounds: magnetite Fe_3O_4 or/and maghemite γ-Fe_2O_3. Thus, these particles, acquiring certain magnetic properties, make it possible to relatively quickly and economically extract the spent sorbent from the treated water by magnetic separation, thereby improving the entire purification process. Meanwhile, in the implementation of the magnetic separation of particles of such coal-sorbent, their magnetic susceptibility plays a key role, information about which in the required quantitative form is still practically absent. However, it is hardly possible to obtain data on the magnetic susceptibility of particles of this sorbent without performing appropriate measurements. These data cannot be sufficiently predictable, even if the method and components of "impregnation" and the magnetic properties of these components are known. In each particular case, such data will be individual due to an arbitrary combination of different magnetically active (magnetite, maghemite), as well as other chemical compounds, elements in the composition of the particles of this sorbent.

The problem of obtaining data on the magnetic susceptibility of iron-containing particles is almost as relevant even when the composition of the particles is specified as known. For example, in the case of using magnetite and/or maghemite particles in biology and medicine. Their magnetically controlled targeted delivery (magnetophoresis), for example, through the blood supply system to a problem member of the body, allows you to create a local source of healing thermal effects on it (eddy currents) or/and directly provide it with the necessary drugs (due to sorption capture and "associated" transport). Known data on the magnetic properties of precisely the particles require clarification, although the material of the particles here (Fe_3O_4, γ-Fe_2O_3 and other magnetically active compounds) is clearly specified.

The same applies to iron-containing particles in so-called magnetic and magnetorheological fluids. They are used as magnetically active media for sealing, lubricating, adjusting the stiffness of a car suspension, heat removal, damping, etc. Despite the fact that the material of the particles is clearly specified here (Fe_3O_4, γ-Fe_2O_3 and other magnetically active compounds), nevertheless, the known data of their magnetic properties require clarification, as was also stated in the previous example.

In addition to the above, the following should also be taken into account. A certain amount of information has already been accumulated in the literature, where we are talking about different samples of dispersed materials (with a dispersed phase of magnetically active particles) and data regarding their magnetic properties, including magnetic susceptibility. At first glance, it may seem that these data can be used to directly judge the magnetic properties of the ("isolated") particles themselves. These data should be treated with caution (especially if one intends to use them as key data in solving the problems of magnetic influence discussed above) for a number of reasons.

First. These data, as a rule, reflect the magnetic properties not of the particles themselves, but of special samples of their substance. In addition, these data are often presented in such a way that it is not entirely clear what kind of object they characterize: the particles themselves or their substance. This uncertainty can misinform the user and lead to serious miscalculations in the implementation of the magnetic impact being undertaken. Indeed, if for the substance of particles the magnitude of the magnetic susceptibility can be calculated in units or hundreds or more, then for a particle, for example, of a spherical shape, when its demagnetizing factor is 1/3, this value is less than three units [11], most often—fractions of a unit.

Second. In the case when there are data on the magnetic properties of samples of dispersed materials (powders, composites, conglomerates, suspensions, colloids, etc.), for the most part there is no such important information on the volume fraction (concentration) of particles in these samples. And this makes the problem of estimating the magnetic properties of the particles themselves uncertain.

Third. More detailed analysis is required for the data obtained by studying the magnetic properties of samples in the form of a suspension or colloid. Thus, under magnetic action on such liquid-dispersed samples, the original structure of the sample material, characterized by a certain dispersion of particles, is violated. The particles here are not deprived of the opportunity to freely move around the volume of the sample, interact with each other—with the formation of chains and aggregates of particles [1, 12]. This leads, as shown in [11] on the basis of data analysis [13], to obtaining conflicting data.

Fourth. The magnetic properties of even seemingly single-sort particles in each particular case can be mutually essentially different. Depending on the method and technological features of obtaining particles—when, as is well known, the size and shape of the particles obtained, their true composition play a role in the manifestation of magnetic properties.

Fifth. Often overlooked is the fundamental fact that data on the magnetic properties of particles or the substance of these particles are a non-linear (for magnetic susceptibility—dependence with extremum) function of the strength of the acting field. Therefore, any unambiguous digital indication of a particular magnetic parameter should be perceived as its purely private (current) value.

Therefore, the issue of control of magnetic properties is relevant, as well as the issue of substantiation of an acceptable control method for this method. For example—the control of magnetic susceptibility, iron-containing (and other magnetically active) particles.

2 Conditions for Using Data on the Magnetic Susceptibility of Dispersed Materials to Determine the Susceptibility of Their Particles

2.1 The Concept of Moderately Rarefied Dispersed Material

Quite complicated are the problems of determining the magnetic susceptibility χ of a single magnetoactive particle. On the other hand, the problem of determining χ based on the control (by traditional methods) of the magnetic susceptibility $\langle\chi\rangle$ of a set of such particles seems to be much more accessible, i.e. a sample of a dispersed, for example powder, material, but specially prepared.

In the initial powder (in the form of a filling) state of a dispersed material, when it is magnetized, magnetically active particles in contact with each other exert a magnetic influence on each other. This means that according to the obtained data on the magnetic parameters of such a material, it is impossible to quantitatively judge the magnetic parameters of just a single particle. If, however, we follow the path of reducing the concentration (volume fraction γ—as a dimensionless parameter) of magnetically active particles in a sample of dispersed material, in particular, by adding the dispersed phase of particles of ground sand to the sample [14], then this ensures rarefaction of the dispersed phase of magnetically active particles and their increasing separation. As a consequence, this leads to a decrease in their mutual magnetic influence up to its almost complete disappearance. Then, according to the data obtained, for example, the magnetic susceptibility $\langle \chi \rangle$ of a sample of this material, one can quite definitely judge the magnetic susceptibility χ of a single (of course, average) magnetically active particle. Given the corresponding value of γ; this approach has been tested for iron-containing impurity particles [8, 9, 11, 14, 15].

The acceptability of this approach can be easily shown by the example of the manifestation of a ponderomotive (magnetic) force, in particular, when implementing the corresponding (ponderomotive) method of controlling magnetic susceptibility. So, taking into account the basic expression (1), written for an isolated small particle, in relation to a system of similar small particles (necessarily mutually sufficiently separated within a dispersed sample of small sizes—such that the values of H and $gradH$ remain practically the same) the resulting ponderomotive force will be their sum:

$$F = \sum F_i = \mu_0 H grad H \cdot \chi \cdot \sum V_i. \tag{2}$$

Regardless of (2), the expression for the ponderomotive force acting as a whole on such a system of small particles (as on a quasi-continuous sample of small volume V) can be written as follows:

$$F = \mu_0 H grad H \cdot \langle \chi \rangle \cdot V. \tag{3}$$

Then from the comparison (2) and (3), taking into account the semantic relation $\sum V_i / V = \gamma$, the expression follows:

$$\chi = \langle \chi \rangle / \gamma, \tag{4}$$

which is suitable for determining the magnetic susceptibility χ of an "isolated" particle. In cases where the magnetic susceptibility $\langle \chi \rangle$ data of a rarefied dispersed (with a dispersed phase of these particles) material is used for this, namely, when the volume fraction γ of the studied magnetically active particles in the studied dispersed sample should not exceed a certain (criteria value [11, 14, 15]) values $\gamma = [\gamma]$.

The values of γ corresponding to these cases, indicating that particles in a sample of a dispersed material can be considered sufficiently "isolated", can be easily established by obtaining an experimental dependence $\langle \chi \rangle$ on γ and evaluating the features of its trend, indicating the presence of two pronounced areas, delimited by the value $\gamma = [\gamma]$. The first of the sections, the initial one, is almost linearly ascending, and the next, the second, is non-linear, more intensively ascending. In this case, the linear form of the first section, when there is a direct proportional relationship between $\langle \chi \rangle$ and γ, i.e. when $\langle \chi \rangle \sim \gamma$, indicates that at $\gamma \leq [\gamma]$, magnetically active particles in the studied sample of dispersed material are sufficiently separated and practically do not experience mutual magnetic influence [11, 14, 15]. Each of these particles can be considered "isolated" here. It has a strength H and a certain value of magnetic susceptibility χ and, therefore, to obtain data χ according to (4), one should use $\langle \chi \rangle$ data only from this section, i.e. for $\gamma \leq [\gamma]$.

To determine the magnetic susceptibility of an "isolated" magnetically active particle χ, one should experimentally find the concentration dependence of the magnetic susceptibility of the dispersed material $\langle \chi \rangle$ of such particles, which is key here, and establish its initial linear section. At least for this purpose, one can confine oneself even to single data $\langle \chi \rangle$, but characterizing already obviously quite rarefied sample. At the same time, it should be borne in mind that in some cases additional steps are required to obtain a dispersed (powder) material. For example, this applies to the case of obtaining a powder of magnetically active iron-containing impurity particles, when it is necessary to specially separate (and accumulate) the corresponding particles from a particular medium. This is achieved by using the recognized polyoperational (according to the number of operations, normalized or established based on the conditions of the experiment) [8, 9] magnetic separation of such particles from a fixed sample of the medium and subsequent obtaining a sample of a dispersed (even powder, which is preferred here) material.

Determining the mass of magnetically active impurity particles isolated from a sample of the medium makes it possible to find a number of important accompanying parameters [8, 9]: concentration and mass fraction in the sample. Comparing such data with the data of traditional chemical control of iron-containing impurities (usually in terms of iron or its oxides), with the studied composition of the isolated particles, it is possible to estimate the fractional part of magnetically active particles in the total composition of iron-containing impurities. This is also possible in the case of obtaining data on the initial and residual concentrations of iron-containing impurities, i.e. when the remaining iron-containing impurities are considered, as it were, to be no longer magnetically active. At the same time, the accuracy of the control of the specified parameters increases if, using the same technique of polyoperational isolation (which is implemented in practice, however, in fact, by a limited number of operations), we use an approach based on the functional extrapolation of such isolation [8, 16]. The approach obliges to obtain and analytically describe asymptotically decreasing (as the sequence number of the operation increases) dependencies of the masses of iron-containing impurities measured after each operation (released from the medium sample) and/or their concentrations (in the medium sample).

2.2 Concentration Dependence of the Magnetic Susceptibility of a Dispersed Material. Initial Line Section. Magnetic Susceptibility of an "Isolated" Particle

Obtaining the dependence of the magnetic susceptibility $\langle \chi \rangle$ of a studied sample of dispersed material, for example, powder, on the volume fraction γ of the studied particles in it. In other words, the determination of data $\langle \chi \rangle$ characterizing a sample with defined dispersity as a whole (as quasi-continuous—within the framework of the so-called effective medium [17–24]) does not cause difficulties. To do this, you can use one of the well-known methods—ballistic [21, 25]. However, this method is effective in measurements with a relatively large (to form the required sample) volume of dispersed material. In more frequent cases, when one has to deal with dispersed samples of small volume, it is preferable to use the ponderomotive method mentioned above (Faraday) [8, 14, 26, 27]. The method makes it possible to calculate $\langle \chi \rangle$ from the known expression for F—here in the form of option (3), from which follows the calculation expression for $\langle \chi \rangle$:

$$\langle \chi \rangle = F \, / \, V \mu_0 H grad H. \tag{5}$$

The method is based on the measurement of the ponderomotive force F acting on the studied sample of small volume V, placed in an inhomogeneity magnetic field, namely at a certain "point" of strength H (averaged within the sample volume).

In [8, 14, 26, 27] this method was developed: in the corresponding magnetometer, the surfaces of the opposing pole pieces are made spherical, which ensures the creation of the required working area (with a stable field inhomogeneity) to accommodate the studied sample of one or another material.

Figure 1, using four different samples of dispersed materials as an example, demonstrates the dependences of the magnetic susceptibility $\langle \chi \rangle$ of each of the samples on the volume fraction γ of magnetically active particles [9, 11, 14, 28]. The dispersed phase of the samples were: particles of natural magnetite, particles of ferroimpurities previously isolated from sugar, semolina, and particles of a carbon sorbent modified (by magnetite and maghemite). The data $\langle \chi \rangle$ was obtained at a magnetic field strength $H = 140$ kA/m for the first three samples and $H = 22$ kA/m for the fourth sample.

Attention is drawn to the remarkable fact that each of these experimental dependences $\langle \chi \rangle$ on γ (Fig. 1) has, as noted above, two sections. These sections are delimited by a criterion, almost identical, value $\gamma = [\gamma] \cong 0.15...0.2$, and the first of these sections (for $\gamma \le [\gamma]$) is expectedly linear: $\langle \chi \rangle \sim \gamma$. This indicates that the magnetically active particles in the sample are already quite separated and their magnetic influence on each other is practically not manifested [9, 11, 14, 15]. This means that there is every reason to follow the corresponding provision of the concept of a moderately rarefied dispersed sample, using data $\langle \chi \rangle$ to determine, according to (4), the values of the magnetic susceptibility of individual particles χ from the specified samples, namely from the first section (for $\gamma \le [\gamma]$). Thus, using relation (4), the

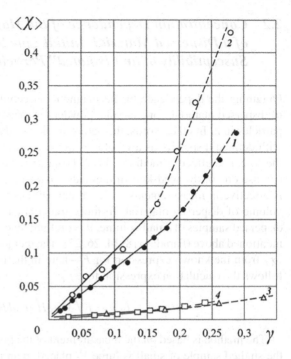

Fig. 1 Concentration
dependences of the magnetic
susceptibility of samples of
dispersed (powder) materials
generalized according to [11,
28]; dispersed phase: 1 (●),
2 (○) and 3 (△)—particles of
natural magnetite,
ferroimpurity particles of
sugar and semolina (all $H =$
140 kA/m), 4 (□)—particles
of modified (inclusions of
magnetite and maghemite)
carbon sorbent ($H = 22$
kA/m)

values of the magnetic susceptibility of particles χ (conditionally secluded) for particles of natural magnetite, ferroimpurity particles isolated from sugar and semolina,
respectively, amounted to χ = 0.75, χ = 1.03 and χ = 0.063 (at $H = 140$ kA/m),
and for modified carbon sorbent particles: χ = 0.095 (at $H = 22$ kA/m).

2.3 Field Dependence of the Magnetic Susceptibility at the Region After Extremum for Dispersed Material (Moderately Rarefied) and "Isolated" Particle

For scientific and practical purposes, detailed (for different values of the field strength
H) information on the magnetic susceptibility of a dispersed sample $\langle \chi \rangle$ and its
reduced value $\langle \chi \rangle/\gamma$ is more in demand. For a moderately rarefied sample ($\gamma \leq [\gamma]$
= 0.15...0.2), the value corresponds to the magnetic susceptibility χ of "isolated"
particles.

Prompt resolution of this issue, i.e. obtaining field dependences $\langle \chi \rangle$ and χ can
be facilitated by the fact that in the most interesting to the user region of H (after
extremum of dependence) [8, 11, 14, 15], these dependences obey power functions
$\langle \chi \rangle \sim H^{-z}$ and $\langle \chi \rangle/\gamma = \chi \sim H^{-z}$—with exponent of power $z = 0.7...0.8$. This is
confirmed by the quasi-linearization of the corresponding dependences of $\langle \chi \rangle$ and
χ on H in logarithmic coordinates.

Such dependences were obtained, in particular, for moderately rarefied powder samples with a dispersed phase of particles of artificial magnetite (Fig. 2a) [8]. Similar (in terms of functional form) dependencies were also obtained for other, moderately rarefied, powder samples (Fig. 2b) with a dispersed phase of ferroimpurity particles. These particles are isolated from condensates of nuclear and thermal power plants, ammonia water and steam for its production, liquid ammonia from the ammonia pipeline, liquid and gaseous ammonia from the production of nitric acid [9, 11]. Similar (by functional form) dependences were also obtained for moderately rarefied powder samples (Fig. 2c) with a dispersed phase of particles of a carbon sorbent modified (with inclusions of magnetite and maghemite) [28].

The indicated power-law form of decreasing dependences $\langle \chi \rangle$ and $\langle \chi \rangle / \gamma = \chi$ on H facilitates obtaining data $\langle \chi \rangle$ and χ for any current (during the implementation of the magnetic effect) value of H. Thus, to determine the values of $\langle \chi \rangle$ and χ at a particular value of the field strength H (in the mentioned region after extremum) it suffices to use the connections:

$$\langle \chi \rangle = \langle \chi \rangle_1 \left(\frac{H_1}{H} \right)^z, \chi = \frac{\langle \chi \rangle_1}{\gamma} \left(\frac{H_1}{H} \right)^z, \qquad (6)$$

and obtained at a certain value of the field strength H_1 with just one (control) value $\langle \chi \rangle_1$. Mandatory for a sample of a rarefied dispersed material (recall that the volume fraction γ of the dispersed phase in it, according to Fig. 1, should not exceed its criterion (transitional) value, i.e. $\gamma \leq [\gamma] \cong 0.15...0.2$). In addition, relations (6) open up the possibility for a simplified solution of the issue related to obtaining the expanded field dependences of the magnetic susceptibility $\langle \chi \rangle$ of the sample of the newly studied dispersed material and its particles χ.

3   Conclusion

It is noted that in many industrial technologies, biotechnologies and to eliminate environmental problems, a targeted magnetic effect is carried out on iron-containing magnetically active particles present in many granular and liquid media. In this regard, attention is drawn to the relevance of the task (to date not yet solved) of obtaining the key information here. These are data on the magnetic susceptibility χ of "isolated" iron-containing particles, which have relatively small dimensions.

An approved concept for determining χ is formulated. It is based on obtaining data on the magnetic susceptibility $\langle \chi \rangle$ of a dispersed material prepared for this sample (with a dispersed phase of iron-containing particles): necessarily moderately rarefied. The volume fraction of γ particles in it should not exceed such a value $\gamma = [\gamma]$ (criterial), when the corresponding mutual separation of particles helps to minimize their magnetic interaction. Then the relation $\chi = \langle \chi \rangle / \gamma$ acceptable for estimating χ becomes valid.

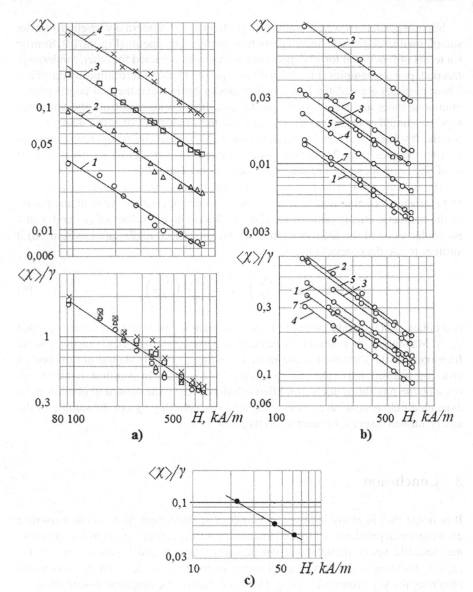

Fig. 2 Field dependences of the magnetic susceptibility $\langle\chi\rangle$ and reduced magnetic susceptibility $\langle\chi\rangle/\gamma = \chi$ (at $\gamma \le [\gamma] \cong 0.2$) generalized according to the data of [8, 11, 28] for samples of dispersed (powder) materials; dispersed phase: **a** particles of artificial magnetite ($1 - \gamma = 0.02$; $2 - \gamma = 0.05$; $3 - \gamma = 0.1$; $4 - \gamma = 0.2$); **b** Ferroimpurity particles isolated from the drainage condensate of a nuclear power plant ($1, \gamma = 0.03$) and return condensate of a thermal power plant ($2, \gamma = 0.15$), ammonia water ($3, \gamma = 0.06$) and steam for its production ($4, \gamma = 0.074$), liquid ammonia from the ammonia pipeline ($5, \gamma = 0.05$), liquid and gaseous ammonia for the production of nitric acid ($6, \gamma = 0.1$ and $7, \gamma = 0.04$); **c** Particles of modified carbon sorbent (with inclusions of magnetite and maghemite) ($\gamma = 0.03$–0.17)

On the example of different powder samples (with magnetically active particles of natural magnetite, ferroimpurity particles isolated from sugar and semolina, as well as with particles of a carbon sorbent modified with magnetite and maghemite), concentration dependences $\langle \chi \rangle$ are demonstrated. On each of them, a linearly ascending initial section, characteristic for determining χ, is traced—up to $\gamma = [\gamma] = 0.15$–0.2.

On the example of different powder samples obtained at $\gamma \leq [\gamma]$ (with particles of artificial magnetite, particles of ferroimpurities isolated from condensates of nuclear and thermal power plants, ammonia water, steam, liquid and gaseous ammonia, particles of a magnetically active sorbent) field dependencies $\langle \chi \rangle$ are demonstrated. Each of them is close to the power $\langle \chi \rangle \sim 1/H^m$ at $m = 0.7...0.8$ in the region of the field strength H (for susceptibility) after extremum. This indicates a similar functional relationship $\chi \sim 1/H^m$ and, moreover, allows one to find the corresponding field dependences $\langle \chi \rangle$ and χ using only the control (for a fixed value of H and, of course, for $\gamma \leq [\gamma]$) value $\langle \chi \rangle$.

Acknowledgements This research was supported by the Ministry of Science and Higher Education of the Russian Federation (project 0706-2020-0024).

References

1. Ngomsik A-F, Bee A, Draye M, Cote G, Cabuil V (2005) Magnetic nano- and microparticles for metal removal and environmental applications: a review. C R Chim 8:963–970
2. Ito D, Nishimura K, Miura O (2009) Removal and recycle of phosphate from treated water of sewage plants with zirconium ferrite adsorbent by high gradient magnetic separation. J Phys Conf Ser 156:012033
3. Sinha S, Ganguly R, De AK (2007) Single magnetic particle dynamics in s microchannel. Phys Fluids 19:117102
4. Mariani G, Fabbri M, Negrini F, Ribani PL (2010) High-Gradient Magnetic Separation of pollutant from wastewaters using permanent magnets. Sep Purif Technol 72:147–155
5. Chen H, Bockenfeld D, Rempfer D, Kaminski MD, Rosengard AJ (2007) Three-dimensional modeling of a portable medical device for magnetic separation of particles from biological fluids. Phys Med Biol 52:5205–5218
6. Baik SK, Ha DW, Ko RK, Kwon JM (2012) Magnetic field analysis of high gradient magnetic separator via finite element analysis. Physica C 480:111–117
7. Wu TH, Mao JH, Wang JT, Wu JY, Xie YB (2009) A new on-line visual ferrograph. Tribol Trans 52:623–631
8. Sandulyak AV (1988) Magnetic and filtration purification of liquids and gases. Chemistry, Moscow, 133 p (in Russ). https://dlib.rsl.ru/viewer/01001440011#?page=136
9. Sandulyak AA, Sandulyak AV, Kiselev DO, Sandulyak DA, Polismakova MN, Ershova VA (2018) About the method of obtaining data of magnetic susceptibility of disperse phase' ferroparticles of powder on concentration and field dependences of its susceptibility. Pribory 11(221):43–51 (in Russ)
10. Vedenyapina MD, Kurmysheva AY, Kryazhev YG, Ershova VA (2021) Magnetic iron-containing carbon materials as sorbents for the removal of pollutants from aquatic media (a review). Solid Fuel Chem 55(5):285–305
11. Sandulyak AA, Sandulyak AV, Ershova V, Pamme N, Ngmason B, Iles A (2017) Definition of a magnetic susceptibility of conglomerates with magnetite particles. Particularities of defining single particle susceptibility. J Magn Magn Mater 441:724–734

12. Yavuz CT, Mayo JT, Yu WW et al (2006) Low-field magnetic separation of monodisperse Fe$_3$O$_4$ nanocrystals. Science 314:964–967
13. Skumiel A, Józefczak A, Hornowski T, Labowski M (2003) The influence of the concentration of ferroparticles in a ferrofluid on its magnetic and acoustic properties. J Phys D Appl Phys 36:3120–3124
14. Sandulyak AA, Polismakova MN, Kiselev DO, Sandulyak DA, Sandulyak AV (2017) On limiting the volume fraction of particles in the disperse sample (for the tasks on controlling their magnetic properties). Fine Chem Technol 12(3):58–64
15. Sandulyak AA, Sandulyak DA, Polismakova MN, Sandulyak AV, Kiselev DO, Ershova VA (2017) Analysis of concentration dependences of magnetic susceptibilities of disperse magnetite-containing media. J Eng Phys Thermophys 90(4):845–850
16. Sandulyak AA, Sandulyak AV, Ershova VA, Polismakova MN, Sandulyak DA (2017) Use of the magnetic test-filter for magnetic control of ferroimpurities of fuels, oils, and other liquids (phenomenological and physical models). J Magn Magn Mater 426:714–720
17. Strečková M, Füzer J, Kobera L et al (2014) A comprehensive study of soft magnetic materials based on FeSi spheres and polymeric resin modified by silica nanorods. Mater Chem Phys 147:649–660
18. Kollár P, Birčáková Z, Vojtek V et al (2015) Dependence of demagnetizing fields in Fe-based composite materials on magnetic particle size and the resin content. J Magn Magn Mater 388:76–81
19. Birčáková Z, Kollár P, Weidenfeller B et al (2015) Reversible and irreversible DC magnetization processes in the frame of magnetic, thermal and electrical properties of Fe-based composite materials. J Alloy Compd 645:283–289
20. Kanhe NS, Kumar A, Yusuf SM et al (2016) Investigation of structural and magnetic properties of thermal plasma-synthesized Fe$_{1-x}$Ni$_x$ alloy nanoparticles. J Alloy Compd 663:30–40
21. Bai K, Casara J, Nair-Kanneganti A et al (2018) Effective magnetic susceptibility of suspensions of ferromagnetic particles. J Appl Phys 124:123901
22. Périgo EA, Weidenfeller B, Kollár P, Füzer J (2018) Past, present, and future of soft magnetic composites. Appl Phys Rev 5:031301
23. Moore RL (2019) Development of a volume fraction scaling function for demagnetization factors in effective media theories of magnetic composites. AIP Adv 9:035107
24. Moore RL (2019) Development and test of concentration scaled demagnetization in effective media theories of magnetic composites. J Appl Phys 125:085101
25. Yurasov AN, Yashin MM (2020) Accounting for the influence of granule size distribution in nanocomposites. Russ Technol J 8(2):59–66 (in Russ)
26. Sandulyak AA, Sandulyak AV, Polismakova MN, Kiselev DO, Ershova VA, Sandulyak DA (2018) The use of spherical pole pieces for performing the Faraday balance method. Instrum Experiment Tech 61(1):123–126
27. Sandulyak AV, Sandulyak AA, Polismakova MN, Sandulyak DA, Ershova VA (2019) The approach to the creation and identification of the positioning zone of the sample in the Faraday magnetometer. J Magn Magn Mater 469:665–673
28. Sandulyak AA, Polismakova MN, Sandulyak DA, Sandulyak AV, Repetunov RA, Kurmysheva AY, Makhiboroda MA (2022) Magnetic susceptibility of powders and magnetic particles (modified inclusions of iron oxides) carbon sorbents. In: International symposium on advanced material and application (ISAMA 2022), Incheon, South Korea (in press)

Data Analysis of the Dynamic Monitoring System for Local Scour of Bridge Piers Based on Acoustic Equipment

Zhi Wei Cui, Hui Sun, and Qing Xuan Qu

Abstract The local scour of bridge piers has always been the main cause of bridge pier instability and bridge damage by water. In terms of local scour monitoring technology, traditional large-scale survey methods have long measurement time intervals, or single-point sensor continuous monitoring methods are poorly representative. In this paper, a dynamic scour monitoring system were introduced based on acoustic equipment. and through the analysis of application case data, the changes and distribution of scour and deposition in the area during the monitoring period and the scour and deposition law in the tides were obtained. Relevant theoretical and model studies were supported by more comprehensive and detailed scientific data through monitoring data and results. This will have a broad application prospect in the related pile erosion protection research field.

Keywords Local scour · Dual-axis sonar · Dynamic monitoring · Data analysis

1 Introduction

The local scour of bridge piers has always been the main cause of bridge pier instability and bridge damage by water [1]. Especially in the estuary and bay area, under the combined action of tidal currents, storm runoff, and offshore waves, the ocean hydrodynamic conditions are complex, and the local seabed of the bridge pier foundation of cross-sea bridges is often affected by three-di mensional flow around the sea bed [2], resulting load-bearing capacity of pier foundations is reduced, which will cause the bridge to be washed out in severe cases. Studying the local scour changes and scour mechanism around the bridge pier is of great significance to ensure the safe operation of the bridge.

Z. W. Cui · H. Sun
Tianjin Survey and Design Institute for Water Transport Engineering Co. Ltd., Tianjin 300456, China

Q. X. Qu (✉)
College of Geodesy and Geomatics, Shandong University of Science and Technology, Qingdao 266590, China
e-mail: quqingxuan@163.com

Most of the traditional scour monitoring is to conduct regular topographic surveys at relatively long intervals [3], which cannot reflect the dynamic evolution process of local scour. Or a sensor installed at a fixed location can obtain terrain change data at a designated point [4], but its representativeness is poor and cannot reflect the local scour. To this end, through extensive technical research, combined with software and hardware development, a dynamic monitoring system based on acoustic equipment was established. The system used sonar equipment fixedly installed on the bridge piers, and the topography of the area around the bridge piers were obtained at fixed time intervals through mechanical rotation. Monitoring data had the characteristics of large coverage area, high precision, good real-time and continuity, which can fully obtain the scouring situation around the bridge pier and timely find the scouring problem of the bridge.

At present, the dynamic monitoring system has been applied to the key piers of a bridge in the southeast coast of China. Based on this, the composition of the system and data processing and analysis methods were introduced in the paper. And the regional scour situation and laws were summarized through 5 months of monitoring data analysis, scientific data support for the evaluation and model study of regional scour were provided by these rules and conclusions, at the same time as a reference for similar operations and research.

2   The Dynamic Scour Monitoring System

2.1   Data Measurement and Acquisition Methods

The dynamic scour monitoring system was built with monitoring sonar as the core, combined with auxiliary equipment such as 4G DTU, sound velocity meter, and self-developed acquisition and control software as shown in Fig. 1. Seabed surface point cloud data was collected by use the monitoring sonar through dual-axis rotation. In order to ensure the monitoring accuracy and coverage of the system, In this system, two sonar devices were installed on both sides of the pier to cover the area around the pier completely. The system operation process is collected sound velocity and attitude data during the scanning and measurement of the sonar equipment synchronously. The main function of the acquisition and control software was to automatically collect the time, correct the monitoring data, and remotely transmit the data to the receiving terminal through 4G DTU. In order to extend the service life of the equipment, the collection interval was set to every 6 h according to the irregular half-day tides at

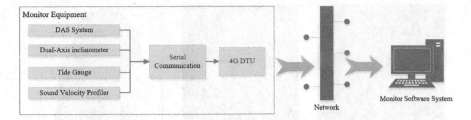

Fig. 1 The dynamic scour monitoring system

the site (that is, 4 sets of monitoring data were collected every day). The monitoring system was installed on the 32# pier of the bridge. The adjacent piers (31#, 33#) were located on the north and south sides of the bridge, and the distance between the piers was 20 m.

2.2 Data Processing and Analysis Methods

Submarine topographic point cloud data was obtained completely by geo-coordinate conversion, attitude calibration, and sound velocity calibration. To facilitate the calculation of the scour distribution and the scour amount, the triangulated irregular network (TIN) was established firstly, then DEM was established through bilinear interpolation based on the TIN. Finally, the normalized grid data was obtained, and the mesh surface was established to realize the discrete digital expression of terrain [5–7], as shown in Fig. 2. To accurately describe the topography around the piers, a spatial resolution of 0.25 m was used to construct a DEM. On this basis, a large amount of long-term monitoring data was summarized, and the overall topographic changes of the region were analyzed through regional analysis and profile analysis, and the law of regional scour and deposition changes during the period was summarized.

3 Data Analysis Process and Results

The scour monitoring period is from December 2020 to April 2021, a total of 5 months and 592 sets of monitoring data are collected. The monitoring data had been summarized and analyzed after correcting, and the scour surplus margin was calculated according to the design scour warning value of the pier at the site of −27.8 m (1985 National Elevation Standard of China, the same below) to determine the safety status of the pier.

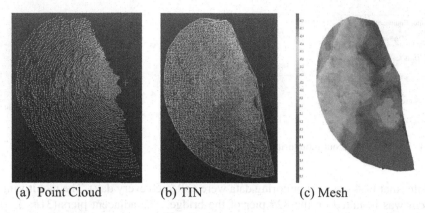

(a) Point Cloud   (b) TIN   (c) Mesh

Fig. 2 DEM construction and rendering diagram

3.1 Overall Regional Situation

The terrain around the monitoring bridge pier 32# fluctuated greatly. There was a raised topography near its roots, the seabed elevation was −16.6 to −19.0 m, the scour surplus margin was 8.8–11.2 m, and its distribution state and seafloor elevation changed little during the period.

There were scour pits near the piers on both sides of the area. For the southern 31# pier, the visible area of the nearby scouring pit was small, and its shape and depth changed little during the period. For the Northern 33# pier, the scour pit nearby actually changed little in December 2020, but there was significant siltation on January 22, 2021, and then turned to scour, and its shape had basically returned to the state at the beginning of monitoring in early February 2021. From February to April 2021, the shape of the scouring pit near 33# pier had little changed. There was scouring and silting on the seabed surface in the pit, and the variation range of scouring and silting was about 0.1–0.5 m. See Table 1 and Figs. 3, 4, 5 for the statistics of scouring pit information during the period. It can be seen from the table that the minimum scour surplus of 33# pier was 4.9 m, which is within the safe range.

Table 1 Information statistics table of 33# pier scouring pit

Time	Form	Elevation of deepest seabed point	Scour surplus
December 2020	28.2 m × 8.5 m 220°	−22.7 m	5.1 m
January 2021	25.4 m × 8.8 m 220°	−22.8 m	5.0 m
February 2021	28.1 m × 9.3 m 220°	−22.7 m	5.1 m
March 2021	29.5 m × 9.4 m 220°	−22.9 m	4.9 m
April 2021	28.7 m × 9.4 m 220°	−22.7 m	5.1 m

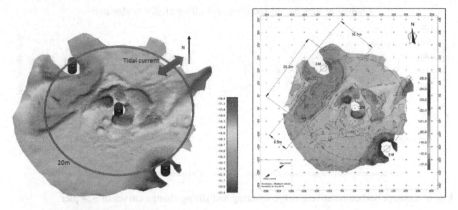

Fig. 3 Three-dimensional topographic map (2020.12.07)

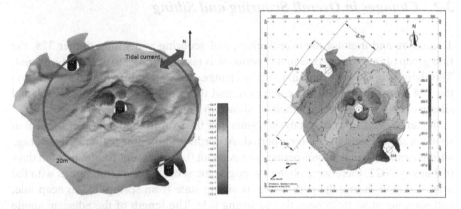

Fig. 4 Three-dimensional topographic map (2021.1.28)

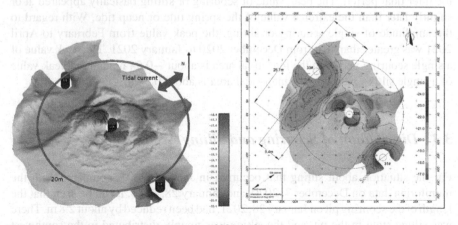

Fig. 5 Three-dimensional topographic map (2021.4.13)

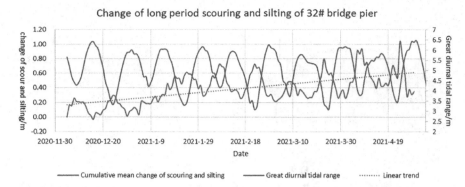

Fig. 6 Long-period cumulative average scouring and silting change curves of 32# pier

3.2 Changes in Overall Scouring and Silting

To analyze and monitor the overall change of scouring and silting in pier 32#, the first group of monitoring data in this period was taken as the statistical starting point. The daily average scouring and silting change of the terrain within 15 m around the 32# pier were statistically calculated, and the cumulative average scouring and silting change curves were plotted, as shown in Fig. 6.

It can be seen from the analysis results that the overall topography of the region shows a silting state over a long period. According to the linear trend, the average silting amplitude in the long-period area is about 0.4 m. Except for the first ten days of January 2021, there were obvious topographic scouring and silting rules with the tidal variation during the period, that is silting state from spring tide to neap tide, and scouring state from neap tide to spring tide. The length of the adjacent single silting and scouring period is the same basically, and it is in good relevance with the inter-tidal period. The peak value of scouring or silting basically appeared at or slightly later than the extreme value of the spring tide or neap tide. With regard to the amplitude of single scouring or silting, the peak value from February to April 2021 was greater than that from December 2020 to January 2021. The peak value of a single scouring period in the inter-tidal area is about −0.54 m, and the peak value of a single silting period in the inter-tidal area is about 0.54 m.

3.3 Distribution of Scouring and Silting Area

For the situation about silting that occurred in January 2021, compared with the monitoring data on December 7, 2020, and January 28, 2021, it can be seen that the length of the scouring pit on January 28, 2021, had been reduced by about 2.8 m. There was silting state in the pit, and the silting was mainly distributed in the southwest side of the pit and the bottom edge of the east side of the pit. The maximum silting

Fig. 7 Distribution of scouring and **silting** changes in the area (2020.12.7–2021.1.28)

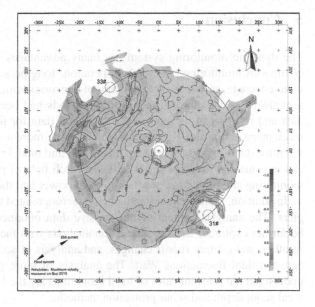

amplitude was 1.6 m. The distribution map of scouring and silting changes is shown in Fig. 7.

4 Summary of the Above Analysis

Through the analysis of monitoring data, it can be seen that the 32# pier scouring surplus is large, and no obvious scouring is found around, so it is in a safe state. There is a large scouring pit nearby the 33# pier, with a depth of about 3.4 m, but the scouring surplus is still 4.9 m, which is also within the safe range. In January 2021, the scour pit nearby 33# pier was once greatly silted. According to the distribution of scouring and silting area, the silting state mainly occurred in the southwest side of the pit and the bottom edge of the east side of the pit, with a maximum silting amplitude of 1.6 m.

In a long period of time, the surrounding of 32# pier shows silting state, indicating that the monitored pier is in good condition. In terms of scouring and silting rules, combined with the analysis of tidal current data, there are spring tide to neap tide silting and neap tide to spring tide scouring rules in the area around the pier during the tidal period.

5 Conclusion

The dynamic monitoring system has many advantages, such as 24 h all-weather continuous monitoring, unattended operation, long period, real-time transmission, wide-coverage, and so on. The problems of low monitoring frequency and poor representativeness of traditional monitoring methods had been solved. The monitoring data and results can be used as the first-hand data for the study of Pier Scouring mechanism and protective methods, and provide more accurate and comprehensive scientific data support for relevant theoretical and model research. It has great application value and broad prospect in the research field of pile foundation local scour monitoring and protection of offshore wind power and the sea crossing bridges.

In addition, there are many influencing factors related to the local scour of bridge piers. The analysis content of the massive data obtained by the system needs to be further explored, such as correlation analysis with factors such as sand content, weather, and estuary runoff changes, and analysis of scouring and silting changes between tidal fluctuations Wait. The analysis of more related factors can further understand the relevant rules of local scour, and lay the foundation for the study of local scour depth and scour protection methods.

Acknowledgements This work was supported by Fundamental Research Fund Project for Central Public Welfare Research Institutes "Research on real-time monitoring technology of local scour based on dual axis scanning sonar" (No. TKS20200302).

References

1. Shunyi W, Li M, Kai W, etc. (2020) Experimental study on local scour of cylindrical pier under different hydraulic conditions[J]. J Disaster Prev Mitig Eng 40(3):425–431
2. Wang L, Yu J, Zhu H, Zeng C, Tang H (2019) A review of the process, influencing factors and numerical modelling of local scour[J]. J Hohai Univ (Nat Sci) 2019(3)
3. Zou J-H, Wang Y-P (2017) The design and implementation of monitoring system for local scour of cross sea bridge piers[J]. J Guangdong Univ Technol 2017(2)
4. Fang S, Shi X, Ni F (2015) Research situation of monitoring technologies for the local scour around piers in Alluvial Rivers[J]. Highw Eng 6:88–95
5. Yin L-M, Jiang X-S, Hu X-L, Pan J-J (2006) Quality control during building DEM[J]. Res Soil Water Conserv 13(3):99–101
6. Jia J, Zhai J, Meng C, etc. (2008) Construction and visualization of submarine DEM based on large number of multibeam data[J]. J GeomatS Sci Technol 25(4):255–259
7. Sun L, Liu Y, Li M (2009) Technique on building DEM of bathymetry using multibeam data[J]. Hydrogr Surv Charting 29(1):39–41
8. Yuan J (2018) Research on underwater DEM based on IMAGENEX DT101 multibeam bathymetry system[J]. GeomatS & Spat Inf Technol 41(4):158–160,163,166

Mobile Robot and Intelligent Control System

Adjustable Magnetic Adsorption Omnidirectional Wall-Climbing Robot for Tank Inspection

Jie Li [ID], Linjie Dong [ID], Mengqian Tian, Chunlei Tu, and Xingsong Wang [ID]

Abstract In-service industrial tanks require periodic inspection and maintenance to ensure safe operation. Wall-climbing robots can effectively replace manual inspection operations, saving time and costs. This paper presents a novel adjustable magnetic adsorption omnidirectional wall-climbing robot for tank inspection. The designed lifting mechanism scheme is a multi-link mechanism, which can adjust the magnetic attraction force of the robot. The magnetic adsorption lifting mechanism provides a stable adsorption force to the robot. The robot can easily disassemble from the working surface. The adsorption distance is analyzed to ensure that the robot and the tank surface without interference. The experimental results demonstrate that the robot has a stable wall-climbing ability. Based on PID controllers and the motion compensation, the stability of horizontal and longitudinal movement is improved. This robot can carry equipment and tools to inspect and maintain on the tank surfaces.

Keywords Climbing robot · Inspection robot · Tank inspection

1 Introduction

Industrial tanks are widely used and require regular inspection and maintenance to ensure safe use. Current manual inspections are time-consuming and labor-intensive, and workers are at high risk. Climbing robots can carry non-destructive testing equipment to replace original manual method. For uncertain radiation risks during the radiographic inspection process, the automated inspection robot can ensure the safety of the inspectors and reduce the risk to a lower level. The inspection robot can execute high-quality, high-precision, and high-efficiency non-destructive inspection of industrial tanks and ensure the safe operation of special equipment in time. The

J. Li
College of Automation, Nanjing University of Posts and Telecommunications, Nanjing, China

J. Li · L. Dong · M. Tian · C. Tu · X. Wang (✉)
School of Mechanical Engineering, Southeast University, Nanjing, China
e-mail: xswang@seu.edu.cn

© The Author(s), under exclusive license to Springer Nature Singapore Pte Ltd. 2023
Y. Ma (ed.), *Advanced Theory and Applications of Engineering Systems Under the Framework of Industry 4.0*, https://doi.org/10.1007/978-981-19-9825-6_6

efficient work of the robot can greatly shorten the inspection period and reduce the economic loss of enterprises.

Climbing robots are a special type of mobile robots, including pipe climbing robots, wall-climbing robots and spherical climbing robots. The climbing mode of the robot can be divided into wheel type, crawler type, multi-legged type and so on. According to purposes and functions, a variety of climbing robots with different structures and driving methods have be designed, such as negative pressure adsorption wall climbing robots [1], electromagnetic adsorption climbing robots[2], electrostatic adsorption wall climbing robot[3], etc. Each form of climbing robots has its own application characteristics.

An omnidirectional mobile wheeled magnetic climbing robot has been developed [4]. The robot used omnidirectional wheels, which have been widely used in mobile robots. The main advantage was that it can greatly improve the mobility of the robot. The omnidirectional wheels were installed on the robot at 120°. The robot can move in arbitrary directions and rotate around its own axis. The chassis with omnidirectional wheels can adapt to different curvature surfaces. A crawler-type climbing robot has been studied [5]. The robot could climb in vertical pipelines, and its crawler was composed of permanent magnet blocks. The independent crawlers could adapt to the change of surface curvature at the turning of pipelines. The robot is equipped with a wireless control system to control the independent movement of two crawlers. Using a wireless camera module, the climb robot could directly detect defects such as cracks in pipelines.

We have proposed a variety of wall-climbing robots [6–8]. Stable adsorption and climbing are basic requirements for wall-climbing robots. In this study, we propose an adjustable magnetic adsorption omnidirectional wall-climbing robot for tank inspec- tion. The designed robot can adsorb on metal walls and move in all directions. The robot with magnetic adsorption lifting mechanisms can change the adsorption force to achieve adsorption and unloading from walls. During the adsorption process, the magnetic adsorption lifting mechanisms can provide stable magnetic adsorption force of the robot. During the unloading process, the magnetic adsorption force can also be easily removed.

2   Robot Design

The wall-climbing robot with magnetic adsorption lifting mechanisms need to meet the following requirements:

- Ensure that adsorption mechanisms can provide stable and reliable adsorption force at the adsorption position.
- Complete the lifting and lowering of permanent magnets through the lifting mechanisms.
- The magnetic adsorption lifting mechanisms can be manually adjusted and lifting action of the mechanisms can be completed easily.

2.1 Mechanism Principle

The designed lifting mechanism scheme is a multi-link mechanism. During robot climbing walls, magnetic adsorption lifting mechanisms can ensure that permanent magnets and the working surface are kept at a certain distance, thereby providing sufficient magnetic adsorption force. When the robot needs to be disassembled from wall surfaces after inspection operation, the lifting handles of the mechanisms can change the rotation position of magnets to reduce magnetic adsorption force.

The schematic diagram of the magnetic adsorption lifting mechanism is shown in Fig. 1. Handle link 1 can be rotated along one revolute pair as a force lever. Upper link 3 is driven to rotate by intermediate link 2. Link 4 drives link 6 to rotate around the bottom point of fixed link 5. The permanent magnet can be rotated and lifted with link 6. After the magnet is rotated, the adsorption force on walls will be reduced, and the robot can be easily removed from the working surface. The maximum rotation angle of the permanent magnet is 28°, and the magnetic adsorption force at this time is weakened by more than 80%.

The link parameters of the lifting mechanism are shown in Table 1. When the robot is in the adsorption state, upper link 3 and link 4 are on the same line. The magnet reaches the lower limit position, and the distance between the magnet and the working surface is 6 mm. When handle link 1 is pushed to the vertical direction, it reaches the upper limit position. The magnet is in the lifting state, and the adsorption force of the robot is greatly weakened. Handle link is designed as a long link, and its lever arm is longer, which is labor-saving.

Fig. 1 Schematic diagram of magnetic adsorption lifting mechanism

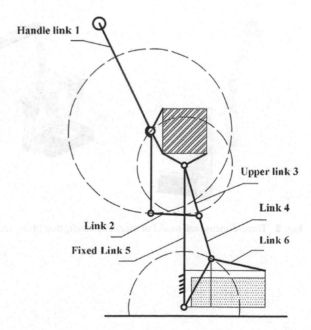

Handle link 1

Upper link 3

Link 4

Link 6

Link 2

Fixed Link 5

Table 1 Link parameters of the lifting mechanism

Links	Parameters
Handle link 1	Upper part 80 mm and lower part 55 mm
Link 2	32.9 mm
Upper link 3	35 mm
Link 4	30.18 mm
Fixed link 5	95.5 mm

2.2 Mechanism Design

After analyzing the principle of the magnetic adsorption lifting mechanism and calculating the geometric size, the magnetic adsorption lifting mechanism is designed and optimized. Figure 2 is the three-dimensional model of the magnetic adsorption lifting mechanism. The whole magnetic adsorption lifting mechanism consists of four groups of single lifting mechanisms. The lifting mechanisms on the same side are connected by one handle, and four permanent magnets are rotated and lifted through the double-sided handles.

In the design of the magnetic adsorption lifting mechanism, miniature bearings are used at rotation points of links to increase the smoothness and stability of revolute pairs. In order to save installation space, the links are all designed in a flat shape. Through the reserved bolt fixing positions, the lifting mechanism can be easily fixed on the robot frame.

Fig. 2 Three-dimensional model of magnetic adsorption lifting mechanism

2.3   Robot Overall Design

Based on the designed magnetic adsorption lifting mechanism, we designed the robot frame, suspension mechanisms, and drive mechanisms. The robot prototype has been built and tested.

Figure 3 shows the 3D model of the adjustable magnetic adsorption omnidirectional wall-climbing robot. The robot is driven by four Mecanum wheels, the maximum load of each wheel is 50 kg, and the diameter of the transmission shaft is 30 mm. The robot frame adopts sectional materials, which reduces the weight of the robot and is easy to build and change. The designed magnetic adsorption lifting mechanisms provide adsorption force, and the robot can stably adsorb and climb on tank surfaces. The distance between four permanent magnets and the bottom surface of Mecanum wheels is 5–8 mm, which can ensure sufficient adsorption force. By pushing the handles on both sides at the same time, the permanent magnets can be rotated and lifted, the robot's magnetic adsorption force will be reduced, and the unloading of the robot from working surfaces can be quickly completed.

Figure 4 shows the dimensions of the climbing robot. The basic size of the robot frame is 400 × 400 mm. The height of the robot is 118 mm. The total dimension of the robot, including suspension mechanisms and Meacanum wheels, reachs 572 × 620 mm. The distance between the front and rear wheels is 540 mm. The diameter of Mecanum wheels is 127 mm and their width is 80 mm.

The actual engineering prototype of the robot is shown in Fig. 5. The magnetic adsorption lifting mechanisms can easily adjust the position of permanent magnets. By applying force to the handles on both sides, all links can be driven, and the permanent magnets are rotated and lifted. The robot can achieve stable adsorption to metal walls and maintain a 6-mm gap between permanent magnets and the working surface. When climbing vertical metal walls, magnetic adsorption mechanisms can meet the operation requirements of the robot and provide enough magnetic adsorption force. Some shortcomings of the robot prototype were found in the preliminary tests. Since the mechanical parts are made of 45# steel, the quality of the robot is 20.8 kg. Lighter materials will be used to reduce the weight of the robot.

Fig. 3 Three-dimensional model of adjustable magnetic adsorption omnidirectional wall-climbing robot

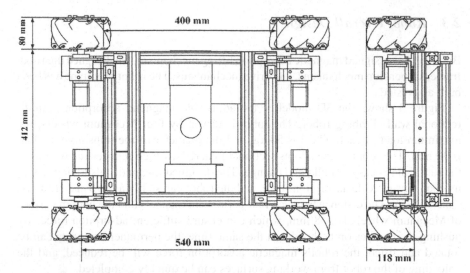

Fig. 4 Dimensions of the climbing robot

Fig. 5 Prototype of the
climbing robot

3 Motion Interference Analysis

Four permanent magnets provide the adsorption force to the climbing robot. The
dimensions of permanent magnets are $100 \times 50 \times 20$ mm. Tank surfaces are cylindrical or spherical, and there may be some weld seams. The permanent magnets need

to maintain suitable gaps with the working surface and not interfere. The interference between the permanent magnets and the working surface will affect the normal movement and adsorption of the robot.

Due to the large span of the permanent magnets and the curved working surface, when the robot is attached to the curved surface in a normal posture, the gaps between the permanent magnets and the curved surface may change from large to small, and even interference may occur. As the radius of tank surfaces changes, the gap between permanent magnets and working surface also changes. When the gap is kept within the range of 4–8 mm, the robot can obtain sufficient adsorption force. This section will analyze the possible interference between the magnet and the curved working surface of different diameters.

When the robot adsorbs and climbs the circular tank surface, the gap between the magnet and the working surface with different diameters (3000 mm, 4000 mm and 5000 mm) is calculated. The permanent magnets are installed on both sides of the robot frame, and the distance between the two magnets on the same side is 540 mm. In order to avoid the magnet adsorption distance being too small, the distance between the magnet center and the working surface is selected as 8.5 mm, and the maximum and minimum distances between the magnet and the working surface are analyzed according to their geometric relationship.

Figure 6 shows a diagram of the distance between the magnet and a curved tank wall with 3000 mm diameter. The wall thickness is 10 mm. When the center of the magnet surface is at 8.5 mm of the arc surface, the following results can be obtained:

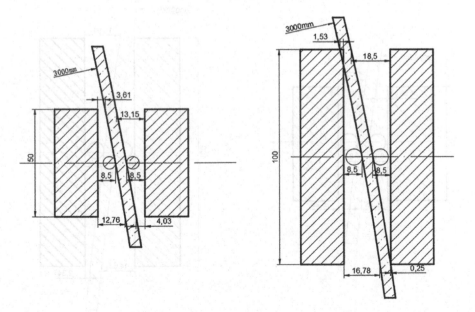

Fig. 6 Interference analysis of robot on a 3000-mm diameter tank surface

(1) When the robot climbs on the inner surface, the minimum distance between the 50 mm side of the magnet and the working surface is 3.61 mm and the maximum distance is 12.76 mm. Larger distances may cause the magnet to not provide enough force. The upper end of the 100 mm side of the magnet has interfered with the curved surface, which has caused the robot to fail to move.

(2) When the robot climbs on the outer surface, the minimum distance between the 50 mm side of the magnet and the working surface is 5.26 mm and the maximum distance is 12.06 mm. The minimum distance between the 100 mm side of the magnet and the curved surface is 2.34 mm and the maximum distance is 15.94 mm.

Figure 7 shows a schematic diagram of the distance between the magnet and a curved tank wall with 4000 mm diameter. The following results can be obtained:

(1) When the robot climbs on the inner surface, the minimum distance between the 50 mm side of the magnet and the working surface is 4.93 mm and the maximum distance is 11.75 mm. The distance between the 100 mm side of the magnet and the working surface is at least 1.04 mm and at most 14.67 mm.

(2) When the robot climbs on the outer surface, the minimum distance between the 50 mm side of the magnet and the working surface is 5.26 mm and the maximum distance is 12.06 mm. The minimum distance between the 100 mm

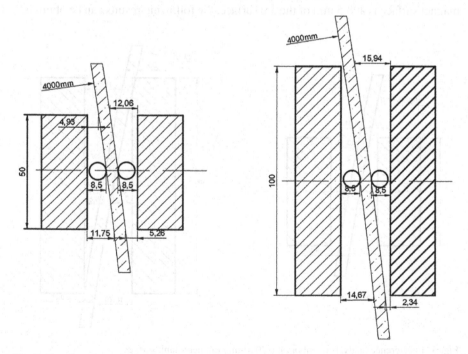

Fig. 7 Interference analysis of robot on a 4000-mm diameter tank surface

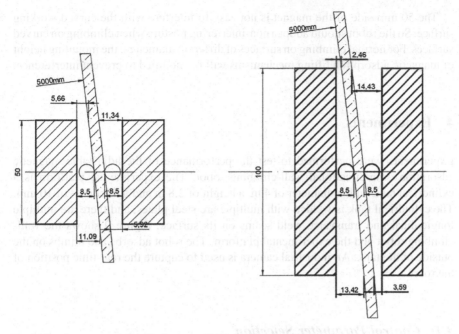

Fig. 8 Interference analysis of robot on a 5000-mm diameter tank surface

side of the magnet and the working surface is 2.34 mm and the maximum is distance 15.94 mm.

Figure 8 shows a schematic diagram of the distance between the magnet and a curved tank wall with 5000 mm diameter. The following results can be obtained:

(1) When the robot climbs on the inner surface, the minimum distance between the 50 mm side of the magnet and the working surface is 5.66 mm and the maximum distance is 11.09 mm. The minimum distance between the 100 mm side of the magnet and the working arc surface is 2.46 mm and the maximum distance is 13.42 mm mm.

(2) When the robot climbs on the outer surface, the minimum distance between the 50 mm side of the magnet and the working surface is 5.92 mm and the maximum distance is 11.34 mm. The minimum distance between the 100 mm side of the magnet and the working surface is 3.59 mm and the maximum is 14.43 mm.

Through the above analysis of the positions of the magnets on the curved surfaces of different diameters, we find that the 100 mm side of the magnet and the 3000-mm diameter working surface have slightly interfered. As the diameter of the working surface increases, the interference will decrease. Excessive distance may lead to unstable adsorption force and bring more instability to the robot. The above analysis is based on the 8.5-mm center distance. As the center distance decreases, the interference problem will become more obvious.

The 50 mm side of the magnet is not easy to interfere with the curved working surface. So the robot should adopt a non-interfering posture when climbing on curved surfaces. For normal climbing on surfaces of different diameters, the mounting height of magnetic adsorption lifting mechanisms will be modified to prevent interference.

4 Experiments

Experiments were performed to test the performance of the adjustable magnetic adsorption omnidirectional wall-climbing robot. The experimental platform is a cylindrical tank with a diameter of 4 m, a height of 2.8 m and a thickness of 10 mm. The cylindrical tank is welded with multiple arc steel plates, and there are multiple longitudinal and transverse weld seams on its surface. Figure 9 shows the wall-climbing robot and the experimental platform. The robot adsorbs and climbs on the outside of the tank. An industrial camera is used to capture the real-time position of the robot.

4.1 Control Parameter Selection

Three PID controllers are used to control the motion of the robot. Appropriate PID control parameters can improve the stability and accuracy of robot motion. The velocity adjustment of the omnidirectional wall-climbing robot includes rotational velocity adjustment, X-direction velocity adjustment (forward motion), and Y-direction velocity adjustment (transverse motion). The three motion velocities of the robot are individually parameterized and tested. When one motion velocity of the robot is tested, the velocities of other two motion will be set to 0.

Fig. 9 The experiment platform and the wall-climbing robot

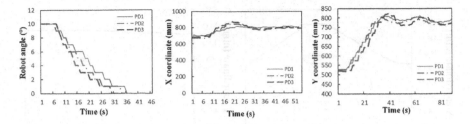

Fig. 10 Adjustment effects of different PID parameters

As shown in Fig. 10, three different sets of PID parameters were tested. Robot angles, X coordinates (vertical direction coordinates), Y coordinates (horizontal direction coordinates) were recorded and compared. PID1 parameters are: KP = 0.3, KI = 0.1, KD = 0.2. PID2 parameters are set as: KP = 0.4, KI = 0.15, KD = 0.3. PID3 parameters are: KP = 0.5, KI = 0.2, KD = 0.4.

In the robot angle adjustment, the robot motion response using PID3 is the fastest. In X-direction and Y-direction velocities adjustment, the robot response of PID2 is faster than that of PID1 and more stable than that of PID3. Therefore, the PID parameters of the robot angle adjustment are selected as: KP = 0.5, KI = 0.2, KD = 0.4. The PID parameters of the X and Y directions adjustment are selected as: KP = 0.4, KI = 0.15, KD = 0.3.

4.2 Climbing Experiments

In the climbing experiments, the robot climbed in the horizontal and vertical directions, respectively. Figure 11 shows X and Y coordinates of the robot when climbing horizontally and vertically. The stability of the robot climbing on tank surface can be judged according to the coordinate changes. During vertical climbing, the robot climbed upward to a 2 m-high position, and then climbed downward to the starting point. The robot takes more time to climb up than it does to climb down (Fig. 11a). The Y coordinate of the robot can be kept stable (Fig. 11b). During the horizontal climbing process, the robot climbed left and right respectively. The Y coordinate of the robot changes steadily (Fig. 11d), while the X coordinate decreases slightly (Fig. 11c). Whether the robot climbs horizontally left or right, the robot has a slight downward movement. Through analysis, due to the influence of gravity, the robot was exerted a downward force due. Sliding down will affect the climbing accuracy of the robot.

In order to solve the downward sliding of the robot during climbing, the X-direction velocity (upward velocity) of the robot is compensated. The experimental results after X-direction velocity compensation are shown in Fig. 12. When the upward velocity compensation is given to the robot, the motion stability of the robot in the X direction is significantly improved, while the motion stability of the robot

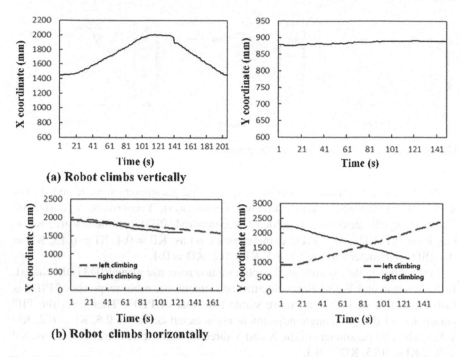

Fig. 11 X and Y coordinates of the robot when climbing horizontally and vertically

in the Y direction has no obvious change. Experimental results indicate X-direction velocity compensation can reduce robot slippage on vertical walls. The robot can climb stably on curved tank surface.

5 Conclusion

In this paper, an adjustable magnetic adsorption omnidirectional wall-climbing robot is introduced. Based on the magnetic adsorption lifting mechanisms, the position of the permanent magnet relative to the metal surface can be changed. The robot can be easily removed from the tank surfaces. Interference analysis of the permanent magnet position and the working surface can determine the optimal adsorption distance and robot posture. The stability of the robot can be improved by X-direction velocity supplementation. The experimental results indicate that the designed robot can climb stably on the tank surface.

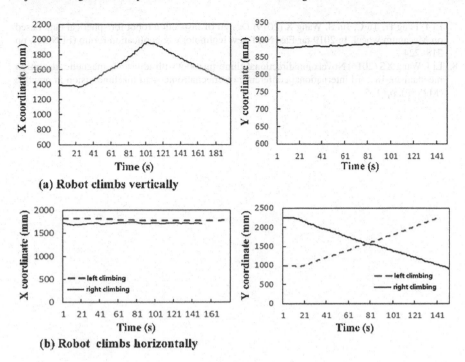

(a) Robot climbs vertically

(b) Robot climbs horizontally

Fig. 12 X-direction velocity compensation of the wall-climbing robot

Acknowledgements This work was supported by the National Key Research and Development Program of China, under Grant SQ2021YFF05002684 and the Science and Technology Plan Project of Jiangsu Province, China, under Grant KJ175933. The corresponding author is Xingsong Wang.

References

1. Ge D, Tang Y, Ma S, Matsuno T, Ren C (2020) A pressing attachment approach for a wall-climbing robot utilizing passive suction cups. Robotics 9:26
2. Zhang W, Zhang W, Sun Z (2021) A reconfigurable soft wall-climbing robot actuated by electromagnet. Int J Adv Rob Syst 18:172988142199228
3. Sabermand V, Ghorbanirezaei S, Hojjat Y (2019) Testing the application of Free Flapping Foils (FFF) as a method to improve adhesion in an electrostatic wall-climbing robot. J Adhes Sci Technol 33:2579–2594
4. Tavakoli M, Lourenço J, Viegas C, Neto P, de Almeida AT (2016) The hybrid OmniClimber robot: Wheel based climbing, arm based plane transition, and switchable magnet adhesion. Mechatronics 36:136–146
5. Nagaya K, Yoshino T, Katayama M, Murakami I, Ando Y (2012) Wireless pip-ing inspection vehicle using magnetic adsorption force. IEEE/ASME Trans Mechatron 17:472–479
6. Li J, Jin S, Wang C, Xue J, Wang X (2020) Weld line recognition and path planning with spherical tank inspection robots. J Field Robot 39:131–152, 2021-10-11 2022

7. Li J, Feng H, Tu C, Jin S, Wang X (2019) Design of inspection robot for spherical tank based on Mecanum wheel. In 2019 far East NDT new technology & application forum (FENDT), pp 218–224
8. Li J, Wang XS (2016) Novel omnidirectional climbing robot with adjustable magnetic adsorption mechanism. In 23rd international conference on mechatronics and machine vision in practice (M2VIP), pp 1–5

Based on Embedded Technology to Realize Control and Validation for Analogous Active Suspension System

San-Shan Hung, Kuo-Wei Lin, and Chi-Chun Hung

Abstract In recent years, the number of electronic control systems on vehicles has been increasing. People's driving comfort, safety, and maneuverability of vehicles have also constituted one of the main considerations when buying a car. Taiwan's roads are highly used, but the quality of road conditions is uneven, which in turn makes the quality of rides declining. Therefore, this research aims at the design of an electronic control system with a analogous active suspension systems for the suspension system of the vehicle to improve the comfort, safety and maneuverability of driving. The vehicle suspension structure used in this research is modified with air springs and proportional valve hydraulic damper to form the characteristics of a analogous active suspension systems. The difference from the active suspension system is that the vehicle does not have an actuator, so the cost it is also low; the electronic control system consists of a vehicle-grade microcontroller, an air spring drive circuit, a damper drive circuit and Various sensors. The micro-controller sends commands to actuate the drive circuit, and then adjusts the air spring or damper to change the elastic coefficient K value and the damping coefficient C value of the suspension system.

Keywords Air spring · Proportional valve hydraulic damper · Analogous active suspension systems

1 Introduction

The traditional sedan suspension system is generally composed of coil shape springs, dampers and connecting rods. The coefficients of the springs and dampers are usually fixed values and cannot be adjusted for specific road surfaces. This research proposes a kind of analogous active suspension systems to replace the passive suspension

S.-S. Hung · K.-W. Lin (✉) · C.-C. Hung
Department of Automatic Control Engineering, Feng-Chia University, Taichung, Taiwan
e-mail: a3738369@gmail.com

S.-S. Hung
e-mail: sshunf@fcu.edu.tw

© The Author(s), under exclusive license to Springer Nature Singapore Pte Ltd. 2023
Y. Ma (ed.), *Advanced Theory and Applications of Engineering Systems Under the Framework of Industry 4.0*, https://doi.org/10.1007/978-981-19-9825-6_7

81

system, such as coil springs and hydraulic dampers changed to pneumatic springs and proportional valve hydraulic dampers with electronic control capabilities, and contains a variety of sensors. In addition to improving the performance of the passive suspension system, this system can also adjust the parameters of the suspension system in real time according to different road conditions, bringing passengers a comfortable, relatively safe and stable riding experience.

2 Research Methods

2.1 Research Flow Chart and Structure

This research will focus on the design and integration of the electronic control unit of the suspension system, which includes the vehicle-grade microcontroller, valve actuator, and sensing unit. The architecture diagram is shown in Fig. 1. The image recognition unit uses the vehicle lens to pre-record the road conditions in front of the vehicle and identify road defects. Finally, the control decision is made to send the control command of the suspension system via CAN Bus, so that the Analogous Active Suspension System can regulate the parameters of the vehicle's shock absorbers in real time.

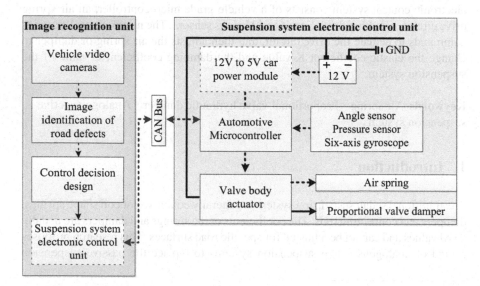

Fig. 1 System architecture

Table 1 Types of suspension systems

Suspension type	Structure and characteristics
Passive suspension system	In the common traditional suspension system, the spring coefficient and damper coefficient are both fixed values
Semi-active suspension system	Independently design an actuator only to adjust the spring coefficient K value or the damper coefficient C value of a fixed suspension system
Active suspension system	Independently design an actuator, and adjust the spring coefficient K value and damper coefficient C value of the suspension system at the same time
Analogous active suspension system	No additional actuators are designed, and the spring coefficient K value and the damper coefficient C value are adjusted through the existing suspension system

2.2 Types of Vehicle Suspension System Structure

The car suspension system is mainly used to reduce the vibration caused by the rugged road when the vehicle is traveling on different road conditions, and at the same time reduce the wear and damage of the mechanical structure of the car body. With the advancement of vehicle manufacturing technology, suspension systems have also developed different types of suspension structures, namely passive, semi-active, active and analogous active suspension systems, as shown in Table 1. Among them, the characteristics of the analogous active suspension systems are (1) there is no additional actuator designed for the active suspension system, (2) the elastic coefficient K and the damper coefficient C are adjustable, so compared with the active suspension system, The cost is lower and can make up for the deficiencies of passive and semi-active suspension systems [1].

2.3 Analogous Active Suspension System Hardware Composition

It is mainly composed of adjustable air spring and proportional valve damper. By increasing or decreasing the internal pressure of the air spring, in addition to changing the spring coefficient K, the ground clearance of the vehicle chassis can also be changed; or when encountering road potholes and high-speed cornering, the spring can provide different degrees for different road conditions. The supporting force can also reduce the wear or damage to the parts when the vibration is transmitted back to the car body. The elastic coefficient of the air spring used in this study exhibits nonlinear characteristics, and its ability to absorb subtle vibrations is better than that of traditional coil springs or hard springs, so it has a significant performance in terms of comfort and reduced wear of parts The proportional valve damper controls the opening and closing of the oil passage inside the damper through the integrated built-in proportional valve. As the current given by the electronic circuit is different, the size of the channel is also different and thus the damping coefficient is changed. Among them, the damping coefficient represents the vibration suppression effect. Generally, the valve channel of the oil circuit inside the damper is a fixed value, and the flow rate of the internal fluid cannot be changed according to different road conditions, thereby affecting the comfort of passengers and the effect of vehicle vibration filtering, resulting in shortened life of vehicle parts. Therefore, by properly adjusting the air spring and proportional valve damper through the design of the electronic circuit, the comfort, controllability and the life of the parts can be improved.

Air Spring Drive Circuit. The air spring is made of rubber material with steel wire, and then a closed air chamber is formed by the upper and lower steel plates, and the elastic non-metallic spring is realized by using the air compressibility and highly elastic rubber material. The folding box-type air spring produced by AirREX, which is selected in this study. Among them, rubber springs of double airbag type and three airbag type are included. The difference is that as the number of layers of the bag body increases, the elastic coefficient increases. In general vehicle weights, the spring elastic coefficient used in the rear axle will be different. Slightly larger than the front axle, the reason is that most of the vehicles have front-end engines and are heavy. Therefore, if the rear axle uses the same spring as the front axle, the rear seat is prone to yaw when the vehicle is cornering, causing the vehicle to be in an unstable state. Therefore, the double-bag spring is installed on the front axle of the vehicle; the shock absorber structure of the rear axle separates the spring and damper, so the three-bag spring is installed on the rear axle.

Controlling the air intake or air release of the spring is through the electronic signal of the microprocessor, which drives the solenoid valve to open and close to control the air inside the air cylinder to introduce the spring or to vent the air inside the spring to the atmosphere to change its spring coefficient K. Among them, the intake and exhaust air are controlled by separate solenoid valves. The four air spring suspension structures of the vehicle have a set of solenoid valves for intake and exhaust, so the shock absorbers of the vehicle can be adjusted independently.

Proportional Valve Hydraulic Damper Drive Circuit. The active suspension system includes dampers in addition to gas springs. The function of the damper is to slow down the mechanical vibration and the device that consumes kinetic energy. When the vehicle encounters a bumpy road while traveling, it will vibrate, and the suspension system will receive the force returned from the road at the first time. At this time, the gas spring will compress and extend with the road, causing the vehicle body to vibrate. The function of the damper is to absorb and slow down the shock caused by the spring, so that the occupant feels comfortable and stable. The principle is to achieve the effect of absorbing shock through the internal fluid and flow of the damper. The internal liquid has a certain viscosity, through which the vibration amplitude can be gradually reduced, reducing the vibration back to the vehicle body. Then adjust the flow through the size of the internal channel and change the damping coefficient C. At present, the modulation principles of adjustable dampers can be roughly divided into two types: (1) changing the width or path of the fluid channel in the chamber through the actuator to lengthen or shorten the compression or rebound time; (2) directly transmits electricity to the coil around the damper chamber and mixes the fluid inside with magnetic particles. By changing the magnetic field, the viscosity coefficient characteristics of the liquid can be changed to change the stretching or compression time. In this study, a proportional solenoid valve is used, and the input current is changed by the microcontroller, so that the channel opened by the path inside the damper becomes larger or smaller.

2.4 Driver Circuit Design

The driving circuit used in this research is mainly used to control the elastic coefficient K of the air spring in the suspension system and the damper coefficient C of the proportional valve hydraulic damper, so that the vehicle has the characteristics of an analogous active suspension systems. The following will give a detailed description of the drive circuit.

Air Spring Drive Circuit. The electronic components used in the air spring include Schmitt triggers, N-MOSFET, flywheel diodes, resistors and capacitors. The circuit is shown in Fig. 2a. The principle of operation is to control the GATE terminal of the MOSFET by the digital signal of the microcontroller to control whether the voltage from the Drain terminal to the Source terminal is turned on and push the solenoid valve to turn on or off. Among them, the flywheel diode plays an important role. When the power is turned off, it can prevent the occurrence of surge voltage and avoid the damage of electronic components [2].

Proportional Valve Hydraulic Damper Drive Circuit. The drive circuit of the proportional valve hydraulic damper is similar to the drive circuit of the air spring. The circuit is shown in Fig. 2b. The difference is that the input signal at the Gate terminal is a PWM signal, belongs to analog signal; in addition, the flywheel diode has more stringent requirements and uses ultra-high speed The switch diode can

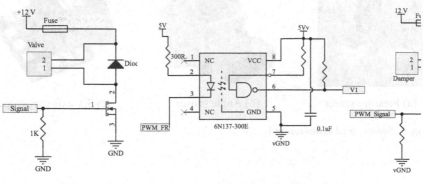

(a) Air spring drive circuit   (b) Damper driver circuit

Fig. 2 Valve body actuator circuit

ensure that the MOSFET can be turned on and off quickly when receiving the PWM signal; the last part is the circuit protection part, in order to avoid the possibility of surges and damage the circuit, so an optocoupler is additionally provided for the damper drive circuit, Isolate the electronic signal of the microcontroller to ensure the stable operation of the system. The proportional valve hydraulic damper used in this study uses a fine-current adjustment method. Its operating principle is that the microcontroller outputs PWM signals to the Gate terminal of the MOSFET, so that the power supplies at the Drain and Source terminals can be quickly turned on and off to achieve current adjustment. The ability to change [3, 4].

2.5   Selection and Description of Sensors

The sensors used in this study include pressure sensor model 92CP8-11, angle sensor model RTY360LVNAX, and six-axis gyroscope model ISM330DLC as shown in Fig. 3. The shock absorbers of the vehicle suspension system are all independently adjustable, so a total of five pressure sensors including air cylinders were used to measure the air pressure of the air spring bladders and four angle sensors to measure the displacement of the four-wheel suspension springs and a six-axis gyroscope to sense the vehicle's attitude.

   The configuration of the sensing circuit is shown in Fig. 4. Among them, the pressure sensor and the angle sensor are connected to the change of the voltage value. In order to reduce the noise, a filter circuit is added to the signal output pin to reduce the excessive voltage value jump; the six-axis gyroscope is an electromechanical system (Micro Electro Mechanical System, MEMS) package form, the related electronic circuits have been integrated into the chip.

(a) Pressure sensor   (b) Angle sensor   (c) Six-axis gyroscope

Fig. 3   Sensors used in this research

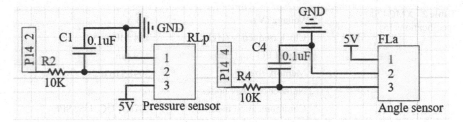

Fig. 4 Circuit configuration of the sensor

2.6 Microcontroller

As the system core of the vehicle electronic control unit, the microcontroller has requirements in terms of performance and computing efficiency and must meet the communication specifications of the automotive AEC-Q100 and 101 series. In this study, the 32-bit automotive chip XMC4700 produced by Infineon was selected. It integrates a lot of units in the chip. In addition to supporting the CAN Bus communication protocol, the I/O pins are up to 114 pins, and the sampling frequency of analog signals can reach 70 ns. It is a very powerful automotive microcontroller, The internal system architecture is shown in Fig. 5, Table 2 shows the basic specifications of the microcontroller [5].

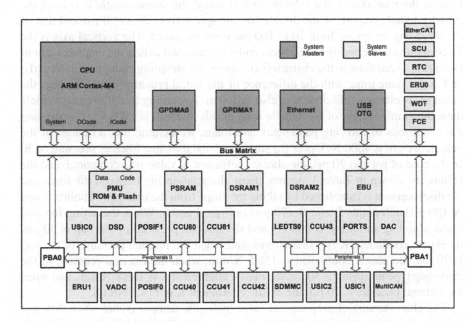

Fig. 5 XMC4700 chip system architecture [5]

Table 2 XMC4700
specifications [5]

Voltage (V)	3.3 V
Clock frequency (Hz)	144 MHz
Digital I/O pin (Pin)	119 pin
Memory (byte)	2048 kbyte
Number of CAN nodes	6
Communication method	UART, I^2C, I^2S, SPI

3 Experimental Results and Analysis

The experimental results of this study include the measurement of the characteristics of the suspension structure (the elastic coefficient K value, the damper coefficient C value measurement), the design and integration of the electronic control unit (including the electronic circuit design, PCB Layout), and the actual vehicle test.

3.1 Characteristic Measurement of Suspension Structure

Figure 6 is the characteristic curve of the double-bag air spring used in the front axle of the vehicle and Fig. 7 is the characteristic curve of the three-bag air spring used in the rear axle of the vehicle. In this study, the characteristic curves of the initial internal pressure of the double-bag air spring from 20 to 90 psi and that of the three-bag air spring from 10 to 100 psi were measured. The vertical axis is the force applied to the spring to be tested, and the horizontal axis is the displacement of the spring. According to the characteristic curve, the air spring changes non-linearly, and at the same time, with the difference of the initial pressure inside the bag, the value of its elastic coefficient K is also different. According to the original vehicle data, the empty weight of the vehicle used in this research is 1405 kg. Considering the conditions of carrying passengers and loads, we assume that a quarter of the vehicle weight is 4000 N. Under the same load and the same bladder pressure of 30 In the case of psi and 70 psi, the elastic coefficient K value was compared, and the results are shown in Table 3. Among them, the relationship between the force and the displacement is established by taking the slope from the changes of points A and A′ (30 psi) and points B and B′ (70 psi) in Figs. 4 and 5. When the acting force of the double-bag air spring is 4000 N and the internal pressure of the bag is 30 psi, its elastic coefficient K is 113.2 N/mm, and when the internal pressure of the bag is 70 psi, the elastic coefficient is 158.5 N/mm; when the internal pressure of the three-bag type air spring is 30 psi, the elastic coefficient K is 135.8 N/mm, and when the internal pressure is 70 psi, the elastic coefficient K is 113.2 N/mm.

The characteristic of proportional valve hydraulic damper is that when different voltage or current values are applied, the size of the internal oil circuit will vary, thus changing the damper C value. Table 4 is a comparison table of the damper

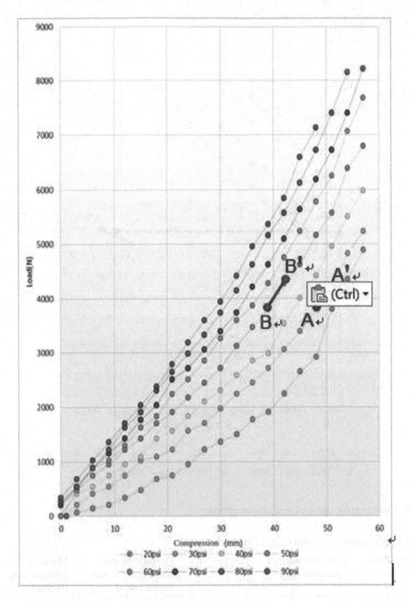

Fig. 6 Characteristic curve of front axle air spring

coefficient C of proportional valve hydraulic dampers and general hydraulic dampers. Among them, the damper coefficient of hydraulic damper is 1,801–2,104 ($\frac{N}{m/s}$); the damper coefficient of compression is 990–1,344 ($\frac{N}{m/s}$), and the damper coefficient of proportional valve hydraulic damper is 1,061–2,326 ($\frac{N}{m/s}$); the damper coefficient of compression is 651–1,151 ($\frac{N}{m/s}$). Among them, when the proportional valve hydraulic

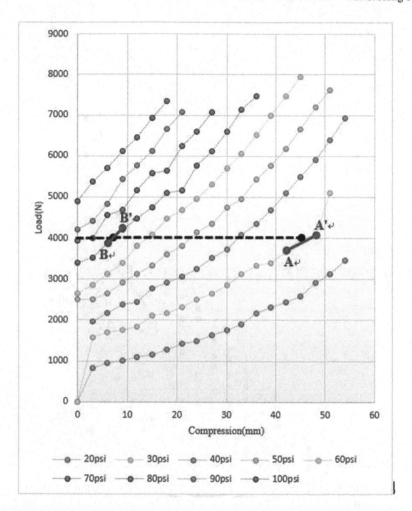

Fig. 7 Characteristic curve of rear axle air spring

Table 3 Air spring coefficient of elasticity K

Front and rear axles	Balloon internal pressure (psi)	Force (N)	The relationship between force and displacement	Elasticity coefficient K (N/mm)
Front axle-Air spring	30	4000	$y = 113.2x - 1633.3$	113.2
	70		$y = 158.5x - 1494.4$	158.5
Rear axle-Air spring	30		$y = 135.8x - 2038$	135.8
	70		$y = 113.2x + 3192.9$	113.2

Table 4 Comparison table of damping coefficient C value between proportional valve damper and passive suspension

Damper state	Damping coefficient C ($\frac{N}{m/s}$)	
	Stretching	Compressed
Front axle hydraulic damper	2,104	990
Rear axle hydraulic damper	1,801	1,344
Proportional valve hydraulic damper energized 0 V	2,112	1,151
Proportional valve hydraulic damper energized 3 V	1,061	654
Proportional valve hydraulic damper energized 4 V	1,341	760
Proportional valve hydraulic damper energized 8 V	2,060	980
Proportional valve hydraulic damper energized 12 V	2,290	831
Proportional valve hydraulic damper energized 14 V	2,326	785

damper is energized at 0 V, the damper coefficient is significantly greater than the energized 3 and 4 V. The reason is that when the damper is energized at 0 V, it will be judged as a damper failure, so the damper coefficient will be slightly greater The state when the power is low. After comparison, the damper coefficient C of the proportional valve damper when it is energized from 4 to 12 V is similar to that of the hydraulic damper when it is extended; the compression is smaller than that of a general hydraulic damper, and the proportional valve hydraulic damper The damper coefficient of the device is similar when it is energized at 0 V.

3.2 Suspension System Controller

During the experiment, we also continued to optimize and improve the control board of the analogous active suspension systems. The circuit board was also changed from a single circuit board to a decentralized structure. The air spring control and the damper control were separated by independent circuits. The board is used for control to avoid damage to other devices when the electronic control system is burned, and to reduce cost losses. At the same time, the size of the circuit board is also greatly reduced. Figure 8 shows the improved circuit board.

3.3 Suspension System Regulation and Response Time

In the control of the vehicle suspension system, the control panel of the electronic control unit issues commands to drive the air springs and dampers. According to the experimental results, the current vehicle can complete the action of the air spring under the output signal of Infineon's automotive chip XMC4700, including air intake

(a) Air spring control circuit   (b) Damper control circuit

Fig. 8 Valve body actuator circuit board

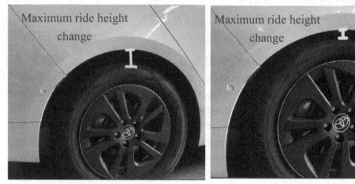

(a) air spring charging   (b) Air spring deflated

Fig. 9 The actuation of air spring intake and its deflation

and air deflation. Figure 9 below shows the difference between the air spring of the vehicle before and after air intake and air deflation.

Regarding the regulation of air springs, this study measured the response time of 0–100 psi and 100–0 psi. Figures 10a and 11a below shows the curve of 100–0 psi; Figs. 10b and 11b shows 0–100 psi. The average inflation of 10 psi falls in the range of 0.5–0.6 s, and the average deflated time of 10 psi falls in the range of 0.85 s.

The response time of the damper in regulation is calculated from the PWM output signal of the controller until the average effect of the MOSFET V_{DS} voltage is reached. Since the PWM output signal varies continuously, the response time in the damper is quite compact, taking about a few milliseconds.

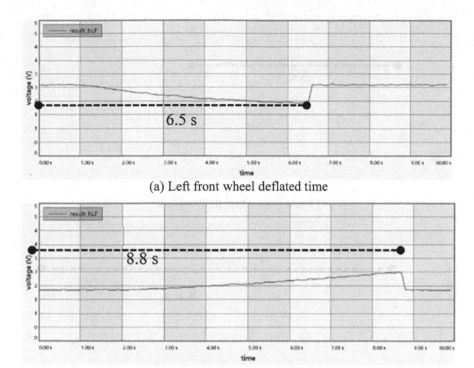

(a) Left front wheel deflated time

(b) Left front wheel inflation time

Fig. 10 Response time of the right front wheel air spring

4 Conclusion

This study measured the structural characteristics of the vehicle before and after the modification, and used the vehicle chip to integrate and design an electronic control unit analogous active suspension systems. Although the rigidity of air springs and proportional valve hydraulic damper may be too large or insufficient, they can be adjusted and appropriately corrected through the intervention of the electronic control unit. At the same time, the electronic control unit also includes sensor monitoring, including the height of the vehicle chassis, the pressure value of the air spring, and the six-axis inertia measurement unit, which can increase the accuracy of controlling the K and C values. Finally, in terms of system response time, the control time of the air spring is slightly longer, and the control time of the damper is shorter. Therefore, subsequent optimizations can be made for the type of air spring capsule and the air ducts for intake and exhaust to shorten the response time of the system.

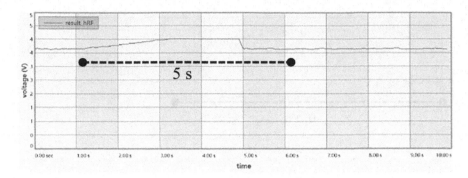

(a) Right front wheel deflated time

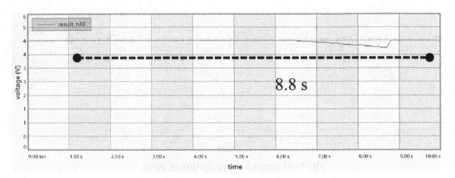

(b) Right front wheel inflation time

Fig. 11 Response time of the left front wheel air spring

Acknowledgements This research was supported by the Ministry of Science and Technology, R.O.C. We appreciate their tremendous support and sponsorship.

References

1. Eligar SS, Banakar RM (2018) A survey on passive, active and semiactive automotive suspension systems and analyzing tradeoffs in design of suspension systems. In: 2018 international conference on recent innovations in electrical, electronics & communication engineering (ICRIEECE), Bhubaneswar, India, 2018, pp 2908–2913. https://doi.org/10.1109/ICRIEECE44171.2018.900 8620
2. Kim H, Lee H (2011) Height and leveling control of automotive air suspension system using sliding mode approach. IEEE Trans Veh Technol Changsha, China 60(5):2027–2041. https://doi.org/10.1109/TVT.2011.2138730
3. Yao K, Hou Z, Zhao X, Tian X (2010) Circuit design of semi-active controller for MR damper. In: 2010 international conference on intelligent computation technology and automation, pp 929–931.https://doi.org/10.1109/ICICTA.2010.367

4. Wang J, Hu H, Wang J, Li Z, Gong Z (2009) Development of an embedded control system for magnetorheological fluid damper under impact load. In: 2009 9th international conference on electronic measurement & instruments, Beijing, China, 2009, pp 3-717–3-722. https://doi.org/10.1109/ICEMI.2009.5274201
5. Infineon Technologies datasheets. https://www.infineon.com. Accessed 2021/5/6

An Exploration of Barrier-Free Driving Education Assistance System Based on HCI Technology Under the Inclusive Design Concept

Ming Yan⬤, Lucia Rosa Elena Rampino⬤, and Giandomenico Caruso⬤

Abstract The massive integration of mobile Internet technology in vehicles led to the launch on the market of many advanced driving assistance systems, with constantly increasing functions and novel modes of interaction. A benchmark on the available systems revealed several usability problems for the drivers, especially those with permanent or temporary impairments. Inclusive design can be achieved by identifying and addressing as many barriers to the Human–Machine Interface (HMI) as possible, focusing on human factors inside the vehicle. It can enable groups with different needs to interact comfortably with the cars and the traffic environment. This paper presents an interactive system based on inclusive design to support users in learning how to use these systems. First, the paper reviews the current research focus and application status of assisted driving existing interactive systems globally. Then, the paper develops and evaluates prototypes through user research, inclusive usage scenarios, and information architecture. Finally, a series of prototype tests verified its usability and provided some interesting insights for future research.

Keywords Interaction design · Inclusive design · Driving education assistance · Human machine interaction · Human-centered design

1 Introduction

In the design of public places, the inclusive design aims at helping different user groups to overcome obstacles and thus support the largest possible number of target users [1]. Currently, inclusive design is also used in transportation, product design, and many other fields [2]. It is a design approach that does not require any adaptation from the user so that mainstream products and services can satisfy as many users as possible.

M. Yan (✉) · L. R. E. Rampino
Department of Design, Politecnico di Milano, Milan, Italy
e-mail: ming.yan@polimi.it

G. Caruso
Department of Mechanical Engineering, Politecnico di Milano, Milan, Italy

With the development of Human–Computer Interaction (HCI) technology, information integration has entered all the aspects of car driving. To ensure acceptance and equality, many diversified barrier-free driving product-service systems have been designed [3].

The current in-car driving interaction methods mainly include: (i) the natural display of the road conditions ahead; (ii) the presentation of assisted driving; (iii) the blend of information and entertainment, and (iv) the integration of the user's mobile devices with the vehicle's system [4]. The human–vehicle interaction in driving teaching activities only focuses on the first two aspects [5]. However, the traditional HCI teaching method in driving could no longer meet the needs of information receiving and processing in the current driving learning process. For example, Traditional teaching is only for exam content, while driving learners also need to understand the vehicle's basic structure; special user groups (i.e., elderly, and disabled people) need more attention, patience, and encouragement; the teaching process of training institutions in different cities is different so it should be standardized. In particular, standardizing the task flow of driving education through effective interaction mode design to accommodate more user groups and improve user satisfaction has become the main task in this field.

2 Related Work

2.1 Current Research Focus

The current automobile driving educational market is very diverse. Due to social and cultural differences, countries have various attitudes towards this field. This paper mainly analyses the Chinese market. In China, in recent years, the Ministry of Public Security has lowered the requirements for disabled people to apply for a driving license, and 279,000 people with physical and hearing disabilities have applied for it. But this only accounts for 1.9% of all driving learners [6].

Users are not familiar with the required gestures and the vehicle parts when learning how to drive, so they have difficulties paying attention to both external traffic and vehicle information. This difficulty severely impacts the disabled and the elderly. Based on this, new interactive ways are gradually entering into driving teaching and developing steadily.

At present, the research on the best interaction modality for driving teaching assistants is mainly divided into two categories: HCI interface and integration model.

From the point of view of the interaction, the traditional HCI interface entails a dashboard and central console physical buttons [7]. In recent years, the touch screen has been gradually extended to in-car spaces. On average, the touch interface of the vehicle console is more intuitive than the physical keys, shortening driving learning. It is a good choice and more convenient and faster [8]. The head-up display (HUD) is also one of them. It uses the principle of optical reflection to project important information on the front windshield [9]. The display height is approximately level with the driver's eyes, and the external scene can be integrated with the display data during driving. Reducing the accident rate between bow and head-up can also alleviate the delay and discomfort caused by the constant eye focus adjustment. However, it has not yet been applied to drivers' education.

On the other hand, the integration model mainly includes voice, somatosensory, and multi-channel interaction [10]. Voice interaction is usually used for daily calls, vehicle navigation, and so on. In the case of a teaching driving experience, voice interaction can correct errors while not interrupting the driver's relevant operations, which relieves teaching burden of coaches. On the other hand, somatosensory interaction is mainly based on gesturing, which requires a specific operation basis, may lead to misjudgment, and is difficult to undo, so it is not suitable for teaching driving. Finally, multi-channel interaction integrates a variety of sensory stimuli such as vision, hearing, and motion sensation; therefore, it could have exciting applications in our field of interest.

In the end, an inclusive driving education system based on HCI technology to accommodate a larger group of users and obtain the satisfaction of both trainees and coaches has yet to be developed.

2.2 Benchmarking of Existing Interactive Systems

The driving test procedures in all provinces and cities of China are unified. If you want to get a driving license, you must pass four tests. The first test evaluates the candidates' theoretical knowledge of the rules of the road. The second is on-site driving: parking and starting on a ramp, side parking, driving on a curve and turning at a right angle. The third test is driving on real roads. The last is a computer-based test on safe driving knowledge. In response to these four tests, the main form of driving education in China is providing a coach. At the same time, interactive systems are relatively limited and mainly divided into web-based applications, mobile applications, and simulators [11].

The method of investigating the Strengths, Weaknesses, Opportunities and Threats (SWOT) of enterprise products and systems is popular among business researchers [12]. In our research, we analyzed seven existing products based on SWOT analysis, on which we collected the following information: main functions, business model, advantages, and disadvantages. Based on this information, a radar chart analysis was carried out in 6 dimensions: user stickiness and ease of use, practicality, interactivity, personalization, and commerciality, as shown in Table 1.

Web-based applications mainly focus on the first and fourth projects of the driving test, describing road rules and safety knowledge standards. They employ video demonstrations with no practical significance, even if they can simulate the driving scene.

Mobile applications combine pictures, text, and video to explain theoretical knowledge. Moreover, they integrate a database of test questions from several

Table 1 Classification and analysis of related products in the Chinese market

Sort	Species	Weakness	Six-dimensional analysis
No.1 Web App	① "Xuechebao" software	Tend to be driving games, emphasis on the pursuit of driving pleasure, lack of actual coaching role There is no teaching part in the demonstration video, only for entertainment	
	② "Moni" software		
No.2 Mobile APP	③ "Hunan Driving Test 2.0" APP	Just focus on the contents of the two theoretical tests	A: User stickiness B: Ease of use C: Practicality D: Interactivity E: Personalization F: Commerciality
	④ "Driving Test Book" APP		
	⑤ "Make driving test easy" APP		
	⑥ "Learn to drive" APP		
No. 3 simulators	⑦ Driving simulation machine	Mainly concentrated in driving schools. Their number is small, and the price is high, making it impossible to be widely used	

provinces and cities. Installing the application, users can browse different databases of questions and make an appointment for the two practical driving tests. Simulators mostly use virtual reality to create a virtual driving training environment. The user interacts with the environment through the machine's operating parts to obtain a driving experience. To a certain extent, this eliminates beginners' fear of driving and standardizes basic driving operations.

3 Research and Design Methods

Our analysis of the market state-of-the-art confirmed that users' specific interactive modalities are the critical factors in the design of interactive systems under the HCI technology. Therefore, a novel design proposal should start from users' research. Firstly, through natural user observation, questionnaire survey, in-depth interviews, the specific needs of users were summarized, and three typical user models were established. Secondly, through inclusive usage scenarios, we outlined the function points that the system should have. Finally, the Micro-Scenario Method (MSM) [13] was used to analyze the system information architecture, determine its content and structure, and define the product specifications to complete the prototype design.

3.1 User Research

Before starting the design activity, it was necessary to understand better the user group's driving learning habits and their level of technological competence. In 2020, the number of cars in China was 270 million, and the total number of people learning to drive was 145 million. Taking Shanghai as an example, there are 202 driving schools, 26,925 coaches, 219,000 students, and 16,997 driving teaching vehicles. The market for driving education is big and has potential for development. Natural observation and questionnaire surveys were used to understand the pain points and requirements of target users. Natural observation is an investigation method where the researcher observes and records the users' behaviors, reactions, and feelings without asking questions. The study was conducted on two driving schools in Shanghai. A total of 527 questionnaires involving 210 among the elderly and people with disabilities were collected.

Summarizing the needs of different user groups, three types of typical users were outlined. Finally, their thoughts, motivations, and attitudes towards assisted driving teaching were researched through in-depth interviews to complete the understanding stage (see Table 2).

3.2 Inclusive Usage Scenarios

The usage scenarios of the product-service system should be based on user needs before prototype development [14]. According to the regulations, people with visual, auditory, and physical impairments can apply for a driver's license if wearing corrective devices does not affect their driving ability. We have applied inclusive design principles to define the system's functions and development details, therefore considering diversified use scenarios. This way, our proposal is also suitable for the disabled, the elderly, and other special groups and improves a satisfying user experience.

We adopted a set of reference charts for inclusive driving learning scenarios. Table 3 summarizes each impairment type and the design countermeasures adopted to guarantee that the overall design is compliant with the six barrier-free standards [15], which is:

1. All text and icons on the interface should be clear and readable
2. The clickable area of different components is far enough
3. Typesetting format conforms to the setting of large font
4. A descriptive link is provided

Table 2 The analysis results of user research

Methods		User requirements	Design point
Natural observation		Driving school venues and vehicles are not regular, and the driving experience is not good The teaching content is not uniform, just for the test	Standard driving school vehicles and venues, unified standards E-learning platform assists in standardized teaching, independent learning
Questionnaire survey		The trainee has no driving experience and is worried about safety hazards Worry that the electronic system cannot communicate with users during teaching, and it is difficult to correct mistakes; under exceptional circumstances, the coach lacks the emergency response-ability The existing teaching model has no test function; therefore, the learner does not understand her level	The voice communication function prompts students when they are wrong; when they are correct, give praise Have automatic protection features such as automatic braking With a mock test function, understand own driving level
In-depth interview	Typical user1-ordinary student	Many people are studying at the same time, and the practice time is uneven Basic knowledge of the car is not explicit, the entire teaching is only for the exam, and it is far from the actual driving	Regulate each person's actual operating time Unified electronic driving teaching system to provide a unified driving experience

(continued)

5. If there are gestures that require a precise operation, make sure that there are other ways to achieve the same function
6. Provide night interface mode corresponding to daytime.

Table 2 (continued)

Methods		User requirements	Design point
	Typical user2-elderly user	The level of driving school coaches is uneven; some have poor driving habits and are likely to mislead students Low acceptance of psychological pressure caused by communication difficulties with the coach, which affects learning progress	The system adopts a voice teaching mode to provide repeated teaching for unique users to avoid communication difficulties
	Typical user3-coach	The coach repeats the tedious teaching work every day, often in a bad mood They are frequently asked questions that are not related to the test, such as vehicle construction	The system can partially replace the coaching job: error correction, repeated explanations, etc.

3.3 Information Architecture

Based on the inclusive usage scenarios, the different user groups that the system can include, and the specific application scenarios were determined. The final product design specifications (PDS) were determined through Kurosu's "micro-scenario method," [13] to obtain the overall system information architecture (see Fig. 1).

The system consists of a voice library and four display interfaces: the vehicle central console, the instrument panel, the head-up display (HUD), and the supporting WeChat applets. Their specific functional modules are shown in Fig. 2. This system aims at realizing the automation and auxiliary functions of intelligent driving teaching. Under certain conditions, students can complete the driving training for passing the practical driving tests (i.e., test 2 and 3) under the system's guidance, without a coach. When they get their driver's license, the system helps users gradually gain confidence with driving on busy roads.

Table 3 Reference standards for inclusive driving learning scenarios, including a detailed classification of impairment types and design countermeasures

Impairment type	Permanent	Situational	Temporary	Design point	Contact point
Visual impairment	Serious eyesight issues (caused by various diseases)	Adverse weather conditions (i.e., heavy rain)	Photophobia after surgery or caused by drugs	Non-visual interaction (i.e., voice interaction and physical buttons.)	Rear-view Mirror/HUD/Central Console/Dashboard/WeChat applet
Limb disorder	Amputation	The operating habits are different	Trauma, joint pain, tension, and anxiety	Voice interaction and camera control methods (such as gesture control and eye control)	Voice system and in-car camera
Hearing/speech impairment	Deafness, speech impediment, autism, etc.	Noisy environment, language barriers, etc.	Transient disorder caused by disease	Replace keyboard input with voice input	HUD/Central Console/WeChat applet
Cognitive impairment	ADHD, depression, autism, etc	Tiredness, distraction, etc	Tension, anxiety	Automatic prompts, voice interaction, ICONS (avoid reading text)	Central control/voice/HUD/WeChat applet

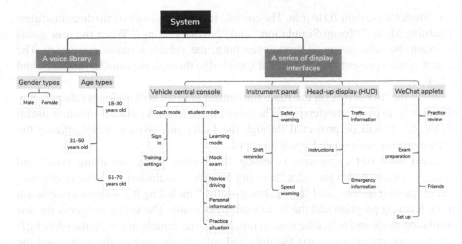

Fig. 1 System information architecture

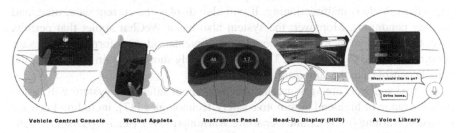

Fig. 2 Summary and display of the prototype framework

3.4 Development and Evaluation of the Prototype

Due to the time and funds limitations, the prototype development concentrated on the vehicle's central console, the instrument panel, and the supporting WeChat applet. The HUD and the voice library will be developed in the research follow-up.

To perform usability testing at the initial development stage, external equipment can be used to replicate ideal future configurations. In our case, an iPad was used as the carrier to connect the system to the vehicle's console and instrument panel through the USB interface [16]. At the same time, a WeChat applet demo was developed. As the technology matures and funds accumulate, the second-generation system will be built into the vehicle. Ideally, it will detect road conditions while driving, correct the driver's wrong operations, and cooperate with car manufacturers to make driving safer.

The usage of the design prototype is as follows: after getting on the car, the coach logs in with a password on the central control screen and enters the teaching content and the training time. In response, the system automatically generates today's training plan. Students use their ID to join. The console interface is divided into three modules: "Learning Mode," "Exam Simulation," and "Novice driving." When the user wants to obtain the information of driving teaching, the vehicle remains stationary. The central control presents the theoretical knowledge through pictures, animations, and videos.

When the user drives on the road, the central console will no longer display any information to avoid interference. Therefore, the relevant visual information useful for the driver will be projected through the HUD and assist in standardizing the driving process with related voice prompts.

When the driver's operation is wrong, the system warns him using visual and auditory stimuli. It also provides "learning progress," including the state of completion of training queries, and the "learning status," including the state of completion of the learning program and the review of errors made. The user can access the test simulation mode and select the item to test. When the vehicle moves to the (driving?) test position, the system starts the quiz and informs the user of the score, and the video shows wrong operations and demonstrates correctly. In addition, the instrument panel has the function of safety prompts. The warning icons are set to flash when the vehicle is malfunctioning. It will also display the current date, time, and weather conditions. Moreover, the system also has a WeChat applet that permits users to check from their mobile the relevant contents and get information on exam appointments and so on. Figures 3 and 4 respectively show the main interface of the vehicle console and the supporting WeChat applet.

As novice drivers are very much concerned about safety, the system will incorporate a range of high-end automotive safety technologies [17], including Forward Collision Warning (FCW), Lane Departure Warning (LDW), Head-Up Display Technology (HUD), blind-spot warning, driver fatigue warning and automatic brake control. Those technologies can help the driver improving driving safety. When

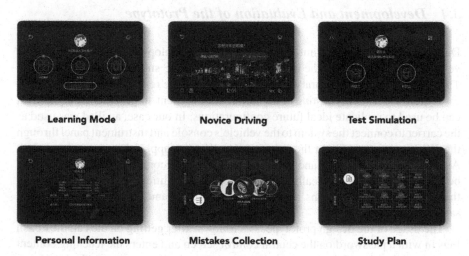

Fig. 3   The main functional interface of the vehicle central console

Fig. 4 The functional interface of the supporting WeChat applet

the vehicle encounters a dangerous situation, the system can automatically generate a braking effect to let the car slow down, helping the driver avoid the standard low-speed driving rear-end accident in urban traffic.

4 Design Evaluation

4.1 Usability Testing

The testing phase of the prototype's usability aimed to find the system's design deficiencies and conduct design evaluation [18]. The test was performed in the Auto Human–Computer Interaction Laboratory of Tongji University. Six subjects among students and teachers were tested. In terms of driving experience, two had more than two years of driving experience, two had just obtained the license, and two were not able to drive a car. In terms of age, three of them were aged 18–30, with strong learning ability; and three were aged 40–60, representing the "mature" user group with weaker receptive ability.

The focus of the test was on the overall evaluation by users of the interactive system for driving education. Since none of the participant had prior experience with intelligent and interactive driving teaching systems, a 10-min video illustrated the system's functioning. When each tester reached a preliminary understanding of the test process and of the interaction modalities, they began the testing process [19], consisting of two parts.

The first part was static, simulating the usage of the central console of the vehicle through a pc tablet. The testers must log in to the student mode when the car is stationery and browse the "introduction to the operation bar". The "Back into the Treasury" project is a project that is often prone to errors in the driving test [20]. The second part of the testing was dynamic. Using the Laboratory driving simulator, the

Table 4 Specific test tasks list and feedback results

Tasks list	Condition	Feedback		
		Interface	Interaction	Functions
1. Log in to the system, select a mode 2. Static operation: learn about the system operating rods 3. Video and audio explanation of related knowledge	1. Provide a dynamic model 2. The driving level of the tested person must meet the requirements	Advantage: 1. Concise and clear, suitable for in-car display 2. The visual appearance of the WeChat applet interface is consistent with the onboard system interface	Advantage: 1. The touch screen, the voice interaction, and the HUD technology are perceived as effective 2. The dangers' warning function improves user satisfaction 3. Convenient integrated service and timely feedback	Advantage: 1. Comprehensive and specific functions, rich in content 2. The security early warning is efficient 3. Timely feedback of questions, improve learning efficiency
4. Dynamic operation: select the "Reverse Storage" item 5. Voice prompt, HUD display information, automatic brake in an emergency 6. Review the WeChat applet		Disadvantages: 1. The central console screen slightly interferes with user groups with low receptivity 2. The functions of the WeChat applet interface are not visible enough	Disadvantages: 1. The voice command is not concise enough 2. The interaction method is a bit cumbersome and not friendly to people with low acceptance	Disadvantages: 1. Worry about technical failures

testers selected this project and learned according to the specific scenes. The detailed tasks list and feedback results see Table 4.

There were two staff members on-site during the test. A staff member introduced the specific operation process to the user before the test and used the camera to record their performance during the whole process. Another staff member asked participants to describe the situations when they had questions and interaction problems and record with pen and paper. The equipment for the experiment and the ongoing simulations are shown in Fig. 5.

4.2 The Tests Result

After completing the above described two tasks, the user was already familiar with the interactive prototype. To understand the problems encountered by the testers

Fig. 5 The test equipment and the scene simulation

while interacting with the system, we performed focus group interviews from which emerged that the testers did not encounter any cognitive barrier or operational difficulties throughout the testing process. On the contrary, using the prototype of the interactive system, they were able to perform all the assigned tasks smoothly. In addition to this, using the method of expert ratings, we selected a pool of experts in related fields at Tongji University, Fudan University, Shandong University in China, and at the University of Sheffield in the UK. The evaluation of the system prototype is based on the ten dimensions of efficiency, accuracy, practicability, standardization, safety, rationality, detail, clarity, user stickiness, and market demand [21]. Experts have given positive feedback on this research. Although they believe that this research is still in the early stages, the direction of this conceptual design is worthy of recognition.

On this basis, an evolved version of the auxiliary interactive prototype will be developed. Our intention is to text this evolved version with impaired users form one side and driving coaches on the other. Since the evaluation process of this study did not conduct targeted research on impaired users, they were only considered as part of the participant selection process to verify the necessity of this concept first. This part will be studied in more detail in the future.

In the future, the system will be also able to adapt its prompting modalities to the specific person and situation, each time adopting a personalized type and degree of sensory stimulation [22].

5 Conclusion

Since different user groups have varied cognition and acceptance of driving education, we conducted an in-depth investigation of users' need and expectations, adopting an inclusive design approach.

Our final aim was to design a barrier-free driving education assistance system based on HCI technology. The system is designed not only to reduce the teaching burden of coaches but also to provide the trainee with a systematic teaching program. In our intention, the designed system can reduce the driving burden of users and is inclusive towards all the users' groups, such as the elderly and the disabled. The main system's interaction modalities were prototyped, and a first round of testing activity was performed.

Based on the results, the usability of the interactive system proposed in this research is positive. There is indeed a certain research space in this direction. It is necessary to further explore the barrier-free driving education assistance system based on human–computer interaction technology under the inclusive design concept. However, since the system is still in the conceptual stage, its feasibility must go through a series of experiments and research, and there is still a long way to go in the future.

References

1. Persson H, Åhman H, Yngling AA, Gulliksen J (2015) Universal design, inclusive design, accessible design, design for all: different concepts—one goal? On the concept of accessibility—historical, methodological and philosophical aspects. Univers Access Inf Soc 14(4):505–526. https://doi.org/10.1007/s10209-014-0358-z
2. Hurtienne J, Horn AM, Langdon PM, Clarkson PJ (2013) Facets of prior experience and the effectiveness of inclusive design. Univers Access Inf Soc 12(3):297–308. https://doi.org/10.1007/s10209-013-0296-1
3. Lyu N, Duan Z, Xie L, Wu C (2017) Driving experience on the effectiveness of advanced driving assistant systems. In: 2017 4th international conference on transportation information and safety, ICTIS 2017—proceedings, pp 987–992. https://doi.org/10.1109/ICTIS.2017.8047889
4. Huang WL, Wang K, Lv Y, Zhu FH (2016) Autonomous vehicles testing methods review. In: IEEE conference on intelligent transportation systems, proceedings, ITSC, pp 163–168. https://doi.org/10.1109/ITSC.2016.7795548
5. Genius D, Apvrille L (2016) Virtual yet precise prototyping: an automotive case study. In: 8th European congress on embedded real time software and systems [Online]. https://hal.archives-ouvertes.fr/hal-01291888
6. Sun X, Jiang Y, Burnett G, Wang Q (2021) Investigating driving styles: a validation study of multidimensional driving styles with British and Chinese drivers. Adv Civ Eng 2021. https://doi.org/10.1155/2021/8831094
7. Flemisch FO, Bengler K, Bubb H, Winner H, Bruder R (2014) Towards cooperative guidance and control of highly automated vehicles: H-Mode and Conduct-by-Wire. Ergonomics 57(3):343–360. https://doi.org/10.1080/00140139.2013.869355
8. May KR, Gable TM, Walker BN (2014) A multimodal air gesture interface for in vehicle menu navigation. In: AutomotiveUI 2014—6th international conference on automotive user

interfaces and interactive vehicular applications, in cooperation with ACM SIGCHI—adjunct proceedings, pp 243–248. https://doi.org/10.1145/2667239.2667280

9. Akyeampong J, Udoka S, Caruso G, Bordegoni M (2014) Evaluation of hydraulic excavator human-machine interface concepts using NASA TLX. Int J Ind Ergon 44(3):374–382. https://doi.org/10.1016/j.ergon.2013.12.002

10. Seif HG, Hu X (2016) Autonomous driving in the iCity—HD maps as a key challenge of the automotive industry. Engineering 2(2):159–162. https://doi.org/10.1016/J.ENG.2016.02.010

11. Martelaro N, Ju W (2017) WoZ way: enabling real-time remote interaction prototyping & observation in on-road vehicles. In: Proceedings of the ACM conference on computer supported cooperative work, CSCW, pp 169–182. https://doi.org/10.1145/2998181.2998293

12. Namugenyi C, Nimmagadda SL, Reiners T (2019) Design of a SWOT analysis model and its evaluation in diverse digital business ecosystem contexts. Procedia Comput Sci 159:1145–1154. https://doi.org/10.1016/j.procs.2019.09.283

13. Ivanova N, Gugleva V, Dobreva M, Pehlivanov I, Stefanov S, Andonova V (2016) We are IntechOpen, the world's leading publisher of Open Access books Built by scientists, for scientists TOP 1 %," Intech, vol i, no. tourism, p 13

14. Narzt W, Mayerhofer S, Weichselbaum O, Pomberger G, Tarkus A, Schumann M (2016) Designing and evaluating barrier-free travel assistance services. In: Lecture notes in computer science (including subseries lecture notes in artificial intelligence and lecture notes in bioinformatics), vol 9752, pp 434–445. https://doi.org/10.1007/978-3-319-39399-5_41

15. Wu Y, Zhao X, Rong J, Zhang Y (2017) How eco-driving training course influences driver behavior and comprehensibility: a driving simulator study. Cogn Technol Work 19(4):731–742. https://doi.org/10.1007/s10111-017-0432-4

16. Vidal M, Bulling A, Gellersen H (2013) "Pursuits: spontaneous interaction with displays based on smooth pursuit eye movement and moving targets. In: UbiComp 2013—proceedings of the 2013 ACM international joint conference on pervasive and ubiquitous computing, 2013, pp 439–448. https://doi.org/10.1145/2493432.2493477

17. Maurer M, Gerdes JC, Lenz B, Winner H (2016) Autonomous driving: technical, legal and social aspects

18. Li J, George C, Ngao A, Holländer K, Mayer S, Butz A (2021) Rear-seat productivity in virtual reality: investigating vr interaction in the confined space of a car. Multimodal Technol Interact 5(4). https://doi.org/10.3390/mti5040015

19. Ogitsu T, Mizoguchi H (2016) A study on driver training on advanced driver assistance systems by using a driving simulator. In: 2015 international conference on connected vehicles and expo, ICCVE 2015—proceedings, pp 352–353. https://doi.org/10.1109/ICCVE.2015.70

20. Lyu N, Deng C, Xie L, Wu C, Duan Z (2019) A field operational test in China: exploring the effect of an advanced driver assistance system on driving performance and braking behavior. Transp Res Part F Traffic Psychol Behav 65:730–747. https://doi.org/10.1016/j.trf.2018.01.003

21. Thurston DL (1991) A formal method for subjective design evaluation with multiple attributes

22. Li S, Zhang J, Wang S, Li P, Liao Y (2018) Ethical and legal dilemma of autonomous vehicles: study on driving decision-making model under the emergency situations of red light-running behaviors. Electronics 7(10). https://doi.org/10.3390/electronics7100264

Design and Analysis of an Orbital Mobile Robot for Cable Tunnel Inspection

Linjie Dong⑩, Jie Li⑩, Mengqian Tian, and Xingsong Wang⑩

Abstract In order to ensure the safe of underground cable tunnels, it is necessary to regularly inspect the lines and equipment in the tunnel. At present, most of the existing tunnel inspection robots can only monitor the environment in the tunnel. The inspection or operation of the lines and equipment in the tunnel needs to be performed manually. In order to make up for the shortcomings of the existing tunnel inspection robots, this paper designs a new orbital mobile inspection robot with a large movement space and gives the overall design of the robot and the detailed structural design of each component. The kinematics model of the robot is established by the D-H method, and the forward kinematics solution, Jacobian matrix and inverse kinematics solution of the robot are deduced. The correctness of the robot kinematics analysis is verified by MATLAB simulation analysis. The research work in this paper lays the foundation for the follow-up research of the robot.

Keywords Tunnel inspection robot · Structural design · Kinematic analysis

1 Introduction

A large number of underground tunnels are built in modern cities for power transmission, communication and underground transportation. In addition, many pipelines are buried underground for the transmission of chemical fuels such as natural gas. The cables in the tunnel and underground pipelines are easily corroded in a dark and humid environment for a long time. Therefore, regular inspections are required for safety [1, 2]. There are many mature pipeline inspection robots in pipeline inspection [3–5], but the inspection of cable tunnels mainly relies on manual inspection, which is inefficient and the sudden failure of equipment in the tunnel will pose a great threat to the personal safety of inspectors in the tunnel.

L. Dong · J. Li · M. Tian · X. Wang (✉)
School of Mechanical Engineering, Southeast University, Nanjing, China
e-mail: xswang@seu.edu.cn

J. Li
College of Automation, Nanjing University of Posts and Telecommunications, Nanjing, China

© The Author(s), under exclusive license to Springer Nature Singapore Pte Ltd. 2023
Y. Ma (ed.), *Advanced Theory and Applications of Engineering Systems Under the Framework of Industry 4.0*, https://doi.org/10.1007/978-981-19-9825-6_9

At present, there is not much research on robots used for cable tunnel inspection. In 2005, Bing Jiang and Alanson P. Sample of the University of Washington developed a set of cable tunnel inspection robots called "Rover" [6], the robot works by crawling along the cable, and the ability to overcome obstacles is weak and the work efficiency is low. In 2009, Shanghai Jiaotong University developed a small cable tunnel inspection robot [7], the robot adopts a crawler-type movement method, and the robot integrates various sensors to extract image information in the tunnel and detect the concentration of various harmful gases. In 2017, China Langchi Xinchuang Technology Co., Ltd. released its new generation of indoor orbital inspection robot A200. This series of robots can realize functions such as temperature measurement and status recognition [8]. In 2019, Zhejiang University designed a set of robots for tunnel inspection [9], the robot carries infrared and visible light cameras to inspect tunnels. These above robots can only monitor the temperature, humidity, and internal environment video in the tunnel through sensors, and none of them have mechanical arms that can approach other equipment and transmission cables in the cable tunnel. At present, the inspection work that needs to be close to the equipment and cables in the tunnel still needs to be carried out manually.

In order to solve these problems and realize the automation of cable tunnel inspection, this paper designs an orbital mobile cable tunnel inspection robot with a mechanical arm. The robot can carry relevant equipment to complete the all-round automatic inspection of the cables and equipment in the cable tunnel.

2   The Structure Design of the Robot

2.1   Internal Environment of the Cable Tunnel

The robot designed in this paper works in the tunnel as shown in Fig. 1. The width of the tunnel is 3.3 m and the height is 2.9 m. Four layers of steel frames are arranged on each side of the tunnel, and the distance between each layer is 0.6 m. The high-voltage cables to be tested are laid on the steel frames, and the mechanical arm is required to reach the position of each cable.

2.2   The Overall Scheme of the Robot

In order to meet the working requirements of the robot, the overall scheme and working schematic diagram of the new inspection robot designed in this paper are shown in Fig. 2. The robot has five degrees of freedom and consists of three parts: a walking system, a lifting system, and a mechanical arm. The walking system and the lifting system realize the movement degrees of freedom in the Z direction and the X direction respectively. In order to achieve a good obstacle avoidance function,

Fig. 1 Dimensional drawing of cable tunnel section (Unit: mm)

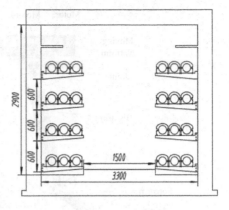

the mechanical arm is designed with three degrees of freedom to achieve flexible movement in the XY plane.

The Mechanical Arm of the Robot. The mechanical arm is driven by rope pulley transmission [10, 11], as shown in Fig. 3. The drive motor is placed on the robot moving platform and the combination of ropes and rollers is used to transmit the drive torque of the motor to each joint, which can effectively reduce the mass and inertia of the moving part of the mechanical arm and reduce power consumption.

The Lifting System of the Robot. The lifting system of the robot adopts a series of parallel links to form a single-degree-of-freedom motion mechanism and is driven by a rope mechanism. The structure has good rigidity and the design of the control system is simple and reliable. Its schematic diagram and structure diagram are shown in Figs. 4 and 5.

The Walking System of the Robot. The movement of the orbital mobile robot relies on the orbit, and it is not easy to tip and shake after being equipped with a lifting mechanism. It can adapt to the complex working environment in the tunnel.

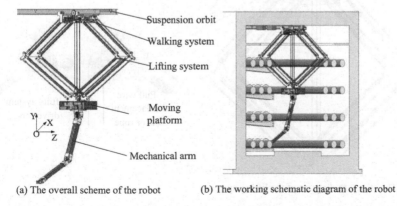

(a) The overall scheme of the robot   (b) The working schematic diagram of the robot

Fig. 2 The overall scheme and working schematic diagram of the robot

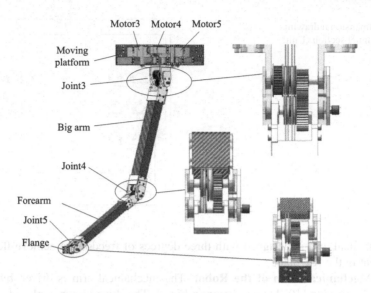

Fig. 3 The structure diagram of the mechanical arm

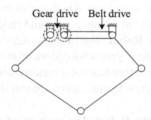

Fig. 4 The schematic diagram of the lifting system

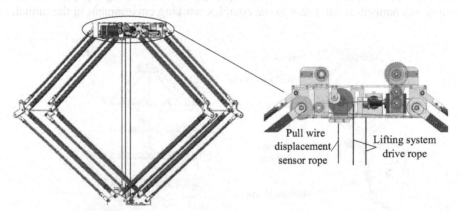

Fig. 5 The structure diagram of the lifting system

Fig. 6 The structure diagram of the walking system

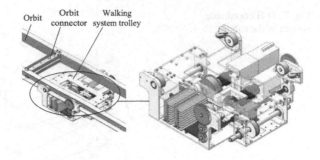

Orbit   Orbit   Walking
               connector   system trolley

Therefore, the walking system of the robot adopts the orbital mobile type, and its internal detailed structure is shown in Fig. 6.

3 Kinematics Analysis of the Robot

3.1 Forward Kinematics Analysis

Establishing the D-H Coordinate System of the Robot. According to the structure characteristics of the robot, the D-H method is used to establish the D-H coordinate system of the robot [12–14], As shown in Fig. 7. First the fixed coordinate system $x_0 y_0 z_0$ is established at the base, and z_0 is taken to coincide with the movement direction of joint 1. The coordinate system $x_1 y_1 z_1$ is established on the robot walking system, z_1 is taken to coincide with the movement direction of joint 2, and x_1 is taken to be the direction of the common perpendicular of the two axes of z_0 and z_1. By analogy, n coordinate system $x_n y_n z_n$ is established on other parts of the robot, z_n is taken to coincide with the axis of the kinematic pair n, and x_n is taken to be the direction of the common perpendicular of the two axes of z_{n-1} and z_n.

According to the D-H coordinate system and D-H transformation rules, the D-H transformation parameter table of the robot can be obtained, as shown in Table 1.

In Table 1, d_i represents the offset along the z_{i-1} axis, θ_i is the torsion angle around the z_{i-1} axis, a_i represents the offset along the x'_{i-1} axis, α_i is the torsion angle around the x'_{i-1}. Where $a_1 = 1650$ mm, $a_3 = 720$ mm, $a_4 = 635$ mm, $a_5 = 50$ mm.

In the initial state of the robot, the initial offset value of the rotation angle θ_i or displacement d_i and the motion range of each joint of the robot are shown in Table 2.

Forward Kinematics Solution of the Robot. The forward kinematics of the robot is to obtain the position and posture of the robot end effector through the angle or displacement of each joint. The forward kinematics is of great significance for the trajectory planning and control of the robot. The following will complete the calculation of the transformation matrix from the coordinate system 5 of the robot end effector to the base coordinate system 0 according to the D-H transformation

Fig. 7 D-H coordinate
system of the robot

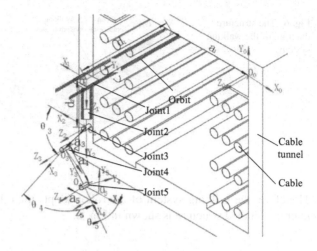

Table 1 D-H transformation parameter table

Coordinate system	d_i /mm	θ_i /°	a_i /mm	α_i /°
0–1	d_1	180	a_1	−90
1–2	d_2	0	0	90
2–3	0	θ_3	a_3	0
3–4	0	θ_4	a_4	0
4–5	0	θ_5	a_5	0

Table 2 Joint position initial offset

Joint variable	d_1/mm	d_2/mm	θ_3/°	θ_4/°	θ_5/°
Initial value	0	525	90	0	0
Range value	–	525–2000	10–170	−80–80	−80–80

parameters between the coordinate systems. The D-H coordinate transformation matrix formula is as follows:

$$_5^0T = {}_1^0T{}_2^1T{}_3^2T{}_4^3T{}_5^4T \tag{1}$$

In formula (1), $_i^{i-1}T$ represents the coordinate transformation matrix that transforms the coordinate r_i in the spatial coordinate system i to the coordinate r_{i-1} in the coordinate system i-1, that is, $r_{i-1} = {}_i^{i-1}Tr_i$. $_i^{i-1}T$ is as follows:

$$i^{-1}_i T = \begin{bmatrix} \cos\theta_i & -\sin\theta_i \cos\alpha_i & \sin\theta_i \sin\alpha_i & a_i \cos\theta_i \\ \sin\theta_i & \cos\theta_i \cos\alpha_i & -\cos\theta_i \sin\alpha_i & a_i \sin\theta_i \\ 0 & \sin\alpha_i & \cos\alpha_i & d_i \\ 0 & 0 & 0 & 1 \end{bmatrix} \quad (2)$$

In formula (2), a_i and d_i refer to Table 1. Where $i = 1, 2, ..., 5$.

Bringing the coordinate transformation parameters of each joint into formula (2), the homogeneous transformation matrix between adjacent coordinate systems of the robot can be obtained. After obtaining the homogeneous transformation matrix between the adjacent coordinate systems, multiply the matrices to obtain the homogeneous transformation matrix of the robot end coordinate system 5 relative to the base coordinate system 0:

$$^0_5T = {}^0_1T{}^1_2T{}^2_3T{}^3_4T{}^4_5T = \begin{bmatrix} n_x & o_x & a_x & p_x \\ n_y & o_y & a_y & p_y \\ n_z & o_z & a_z & p_z \\ 0 & 0 & 0 & 1 \end{bmatrix} \quad (3)$$

Bringing the transformation matrix into formula (3), the expression of each parameter can be obtained as:

$$\begin{cases} n_x = s_5(c_3s_4 + c_4s_3) - c_5(c_3c_4 - s_3s_4) = -c_{345} \\ n_y = -c_5(c_3s_4 + c_4s_3) - s_5c_3c_4 - s_3s_4) = -s_{345} \\ n_z = 0 \end{cases} \quad (4)$$

$$\begin{cases} o_x = c_5(c_3s_4 + c_4s_3) + s_5(c_3c_4 - s_3s_4) = s_{345} \\ o_y = s_5(c_3s_4 + c_4s_3) - c_5(c_3c_4 - s_3s_4) = -c_{345} \\ o_z = 0 \end{cases} \quad (5)$$

$$\begin{cases} a_x = 0 \\ a_y = 0 \\ a_z = 1 \end{cases} \quad (6)$$

$$\begin{cases} p_x = a_5s_5(c_3s_4 + c_4s_3) - a_3c_3 - a_5c_5(c_3c_4 - s_3s_4) - a_1 - a_4c_3c_4 + a_4s_3s_4 \\ \quad = -a_3c_3 - a_5c_{345} - a_1 - a_4c_{34} \\ p_y = -d_2 - a_3s_3 - a_5c_5(c_3s_4 + c_4s_3) - a_5s_5(c_3c_4 - s_3s_4) - a_4c_3s_4 - a_4c_4s_3 \\ \quad = -d_2 - a_3s_3 - a_5s_{345} - a_4s_{34} \\ p_z = d_1 \end{cases} \quad (7)$$

where $s_i = \sin\theta_i$, $s_{ij} = \sin(\theta_i + \theta_j)$, $s_{ijk} = \sin(\theta_i + \theta_j + \theta_k)$, $c_i = \cos\theta_i$, $c_{ij} = \cos(\theta_i + \theta_j)$, $c_{ijk} = \cos(\theta_i + \theta_j + \theta_k)$.

T_{50} is the forward kinematics solution of the robot. In formula (3),n_x, n_y, n_z; o_x, o_y, o_z; a_x, a_y, a_z are the unit vector representation of coordinate axes x_5, y_5, z_5 in the 0 coordinate system, reflecting the posture of the 5 coordinate system relative to the 0 coordinate system; p_x, p_y, p_z represent the position vector of the origin of the 5 coordinate system in the 0 coordinate system.

3.2 Inverse Kinematics Analysis

The inverse kinematics of the robot is the problem of solving nonlinear equations [15], and inverse kinematics solutions may have infinite groups. In this paper, based on Jacobian matrix and differential kinematics, the "Newton–Raphson iteration method" is used to solve the inverse kinematics of the robot. The inverse kinematic solution obtained by this method has only a unique solution, which is equivalent to the final solution determined according to the principle of minimum energy.

Build the Jacobian Matrix of the Robot. The Jacobian matrix of the robot is established by the vector-differentiation method, its specific form is as follows:

$$
J = \begin{bmatrix} J_{v1} & \cdots & J_{v5} \\ J_{\omega1} & \cdots & J_{\omega5} \end{bmatrix} = \begin{bmatrix} 0 & 0 & a_3s_3 + a_5s_{345} + a_4s_{34} & a_5s_{345} + a_4s_{34} & a_5s_{345} \\ 0 & -1 & -a_3c_3 - a_5c_{345} - a_4c_{34} & -a_5c_{345} - a_4c_{34} & -a_5c_{345} \\ 1 & 0 & 0 & 0 & 0 \\ 0 & 0 & 0 & 0 & 0 \\ 0 & 0 & 0 & 0 & 0 \\ 0 & 0 & 1 & 1 & 1 \end{bmatrix}
$$

(8)

Inverse Kinematics Solution of the Robot. The differential motion of the robot joint coordinate system can be described by $D = [dx, dy, dz, \delta x, \delta y, \delta z]^T$, where dx, dy and dz represent differential translation, and $\delta x, \delta y$ and δz represent differential rotation. The relationship between differential motion and generalized velocity is as follows:

$$
\begin{bmatrix} v \\ \omega \end{bmatrix} = \lim_{\Delta t \to 0} \frac{1}{\Delta t} \begin{bmatrix} d \\ \delta \end{bmatrix} = \lim_{\Delta t \to 0} \frac{1}{\Delta t} D
$$

(9)

The generalized velocities in Cartesian space and joint velocities in joint space can be converted to each other by Jacobian matrix. We can get the following formula:

$$
\begin{bmatrix} v \\ \omega \end{bmatrix} = J \frac{dq}{dt} = \lim_{\Delta t \to 0} \frac{1}{\Delta t} D
$$

(10)

By formula (10) we can get the following formula:

$$dq = J^{-1}D \tag{11}$$

The differential motion vector D of the robot can be obtained by the differential motion operator. The differential motion operator of the robot is as follows:

$$\Delta = \begin{bmatrix} 0 & -\delta z & \delta y & dx \\ \delta z & 0 & -\delta x & dy \\ -\delta y & \delta x & 0 & dz \\ 0 & 0 & 0 & 0 \end{bmatrix} \tag{12}$$

The differential motion of the robot can be expressed as the current pose T_{cur} moving to the target pose T_{end}, and the relationship between them is as follows:

$$\Delta = T_{cur}^{-1} T_{end} - I \tag{13}$$

According to formula (13), the values of $dx, dy, dz, \delta x, \delta y$ and δz can be obtained, and the differential motion vector D of the robot can be obtained.

When the coefficient matrix J is a singular matrix, J^{-1} does not exist. At this case, the generalized inverse of the matrix needs to be obtained. In this paper, Householder's SVD decomposition method is used to find the generalized inverse matrix J^{+}, and we can get the following formula:

$$J^{+} = V \Sigma^{-1} U^{H} \tag{14}$$

$$dq = J^{+}D = V \Sigma^{-1} U^{H} D \tag{15}$$

Formula (15) is the Newton–Raphson iterative equation for the inverse kinematics solution of the Robot. When the number of iterations is less than N, and $T_{q+dq} - T_{end} < \varepsilon$, end the iteration and return the joint position $q + dq$ corresponding to the robot target pose T_{end}. When the number of iterations is greater than N, the iteration ends and no suitable solution is found. N is the set number of iterations.

4 Kinematics Simulation Analysis of the Robot

The robot kinematics model is established in the Robots Toolbox module of MATLAB. The simulation results in the Robots Toolbox are compared with the calculation results of the robot kinematics model established in this paper to verify the correctness of the robot kinematics model established and the correctness of the kinematics forward and inverse solutions.

4.1 Forward Kinematics Solution Verification

The joint parameters of the robot at the initial position are set as: $d_1 = 0$ m, $d_2 = 1$ m, $\theta_3 = 60°, \theta_4 = 60°, \theta_5 = 60°$. The joint parameters of the robot at the end position are set as: $d_1 = 2$ m, $d_2 = 1.6$ m, $\theta_3 = 40°, \theta_4 = 40°, \theta_5 = 40°$. The position coordinates of the flange at the end of the robot are solved through the robot kinematics model established in the Robots Toolbox, as shown in Fig. 8.

It can be seen from Fig. 8 that at the initial position, the coordinate of the end effector in the 0 coordinate system is $(-1.643$ m, -2.173 m, 0 m). Bringing $d_1 = 0$ m, $d_2 = 1$ m, $\theta_3 = 60°$, $\theta_4 = 60°$, $\theta_5 = 60°$ into formulas (4)–(7), we can get the following formula:

$$T_{50} = \begin{bmatrix} n_x & o_x & a_x & p_x \\ n_y & o_y & a_y & p_y \\ n_z & o_z & a_z & p_z \\ 0 & 0 & 0 & 1 \end{bmatrix} = \begin{bmatrix} 1 & 0 & 0 & -1.643 \\ 0 & 1 & 0 & -2.173 \\ 0 & 0 & 1 & 0 \\ 0 & 0 & 0 & 1 \end{bmatrix} \tag{16}$$

It can be seen from Fig. 8 that at the end position, the coordinate of the end effector in the 0 coordinate system is $(-2.287$ m, -2.731 m, 2.000 m). Bringing $d_1 = 2$ m, $d_2 = 1.6$ m, $\theta_3 = 40°$, $\theta_4 = 40°$, $\theta_5 = 40°$ into formulas (4)–(7), we can get the following formula:

$$T_{50} = \begin{bmatrix} n_x & o_x & a_x & p_x \\ n_y & o_y & a_y & p_y \\ n_z & o_z & a_z & p_z \\ 0 & 0 & 0 & 1 \end{bmatrix} = \begin{bmatrix} 0.5 & 0.866 & 0 & -2.287 \\ -0.866 & 0.5 & 0 & -2.731 \\ 0 & 0 & 1 & 2 \\ 0 & 0 & 0 & 1 \end{bmatrix} \tag{17}$$

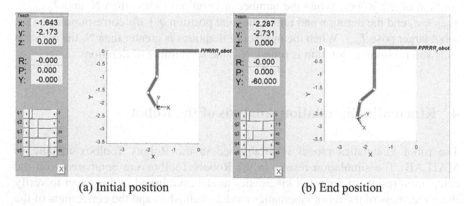

(a) Initial position   (b) End position

Fig. 8 The solver results of Robot Toolbox

The calculation results are completely consistent with the kinematics simulation results in the MATLAB Robot Toolbox, indicating that the established robot kinematics model and the obtained robot forward kinematics solution are correct.

4.2 Inverse Kinematics Solution Verification

The joint parameters of the robot at the initial position are set as: $d_1 = 0\,\text{m}$, $d_2 = 1.3\,\text{m}$, $\theta_3 = 50°$, $\theta_4 = 50°$, $\theta_5 = 50°$. Bringing them into formulas (4)–(7), we can get the forward kinematics solution.

$$T_{cur} = \begin{bmatrix} n_x & o_x & a_x & p_x \\ n_y & o_y & a_y & p_y \\ n_z & o_z & a_z & p_z \\ 0 & 0 & 0 & 1 \end{bmatrix} = \begin{bmatrix} 0.866025 & 0.500000 & 0 & -1.959239 \\ -0.500000 & 0.866025 & 0 & -2.501905 \\ 0 & 0 & 1 & 0 \\ 0 & 0 & 0 & 1 \end{bmatrix} \tag{18}$$

Assuming that at the end position, the pose matrix of the robot is as follows:

$$T_{end} = \begin{bmatrix} n_x & o_x & a_x & p_x \\ n_y & o_y & a_y & p_y \\ n_z & o_z & a_z & p_z \\ 0 & 0 & 0 & 1 \end{bmatrix} = \begin{bmatrix} 0.913545 & 0.406737 & 0 & -1.893978 \\ -0.406737 & 0.913545 & 0 & -2.803842 \\ 0 & 0 & 1 & 2 \\ 0 & 0 & 0 & 1 \end{bmatrix} \tag{19}$$

Bringing formula (18) and (19) into the inverse solution iterative equation, the position q of each joint at the end position can be obtained. The iterative process is shown in Table 3.

Table 3 Joint position iteration process

Number of iterations (N)	d_1/m	d_2/m	$\theta_3/°$	$\theta_4/°$	$\theta_5/°$
1		2.0000	1.4736	59.8071	50.3139	45.8680
2		2.0000	1.6200	56.0391	48.7683	51.1926
3		2.0000	1.6100	52.9333	48.8486	54.2181
4		2.0000	1.5838	52.9818	49.1420	53.8762
5		2.0000	1.5806	53.4932	49.1905	53.3163
6		2.0000	1.5847	53.5738	49.1472	53.2790
7		2.0000	1.5859	53.4982	49.1318	53.3700
8		2.0000	1.5854	53.4712	49.1366	53.3922
9		2.0000	1.5851	53.4796	49.1401	53.3803
10		2.0000	1.5851	53.4857	49.1399	53.3744
11		2.0000	1.5852	53.4853	49.1392	53.3755

Bring the results of the 11th iteration $d_1 = 2.0000$ m, $d_2 = 1.5852$ m, $\theta_3 = 53.4853°$, $\theta_4 = 49.1392°$, $\theta_5 = 53.3755°$ into formulas (4)–(7), we can get the following formula:

$$T_{end}' = T_{q+dq} = \begin{bmatrix} 0.913545 & 0.406737 & 0 & -1.893958 \\ -0.406737 & 0.913545 & 0 & -2.803852 \\ 0 & 0 & 1 & 2 \\ 0 & 0 & 0 & 1 \end{bmatrix} \tag{20}$$

The error between the calculated result of the robot end flange pose and the preset pose is as follows:

$$T_{end}' - T_{end} = \begin{bmatrix} 0 & 0 & 0 & 0.000037 \\ 0 & 0 & 0 & 0.000046 \\ 0 & 0 & 0 & 0 \\ 0 & 0 & 0 & 0 \end{bmatrix} \tag{21}$$

It can be seen from formula (21) that the error between the calculated result of the robot end flange pose and the preset pose is small and meets the required range, that is, the inverse kinematics solution of the robot is correct and feasible.

5 Conclusions

According to the current problems in cable tunnel inspection, this paper designs a five-degree-of-freedom orbital mobile inspection robot, which can carry relevant equipment to complete all-round automatic inspection of cables and equipment in cable tunnels. The robot coordinate system is established by the D-H method, the kinematic equations of the robot are analyzed, and the forward kinematics solution, Jacobian matrix and inverse kinematics solution of the robot are obtained. The robot kinematics simulation analysis is carried out through MATLAB, and it is verified that the established robot kinematics model and kinematics forward and inverse solutions are correct. The work done in this paper lays a theoretical foundation for robot control and subsequent research.

Acknowledgements This work was supported by the National Key Research and Development Program of China, under Grant SQ2021YFF05002684 and the Science and Technology Plan Project of Jiangsu Province, China, under Grant BE2022030776. The corresponding author is Xingsong Wang.

References

1. Hui Z, Qian R (2018) Mechanism design and track optimization simulation of a new type of cable tunnel inspection robot. Mech Des Manuf Eng 47(03):61–65
2. Cui W (2017) Structural design and motion control of tunnel inspection robot. Master, Shenyang Aerospace University
3. Prasad EN, Kannan M, Azarudeen A, Karuppasamy N (2012) Defect identification in pipe lines using pipe inspection robot. Int J Mech Eng Robot Res 1(2):20–31
4. Kusunose K, Akagi T, Dohta S, Kobayashi W, Nakagawa K (2019) Development of pipe holding mechanism and bending unit using extension type flexible actuator for flexible pipe inspection robot. Int J Mech Eng Robot Res 8(1):129–134. https://doi.org/10.18178/ijmerr.8.1.129-134
5. Hayashi K, Akagi T, Dohta S, Kobayashi W, Shinohara T, Kusunose K, Aliff M (2020) Improvement of pipe holding mechanism and inchworm type flexible pipe inspection robot. Int J Mech Eng Robot Res 9(6):894–899. https://doi.org/10.18178/ijmerr.9.6.894-899
6. Jiang B, Sample AP, Wistort RM, Mamishev AV (2005) Autonomous robotic monitoring of underground cable systems. In: 12th international conference on advanced robotics, Seattle, WA, pp 673–679
7. Fu Z, Chen Z, Zheng C, Zhao Y (2008) A cable-tunnel inspecting robot for dangerous environment. Int J Adv Rob Syst 5(3):243–248
8. Xian K (2018) Design of control system for rail-type inspection robot in substation. Master, Southwest Jiaotong University (2018)
9. Lv X (2019) Research on suspended track-type intelligent inspection robot. Master, Zhejiang University
10. Kirchhoff J, von Stryk O (2018) Velocity estimation for ultralightweight tendon-driven series elastic robots. IEEE Robot Automat Lett 3(2):664–671
11. Kim Y (2017) Anthropomorphic low-inertia high-stiffness manipulator for high-speed safe interaction. IEEE Trans Rob 33(6):1358–1374
12. Liu F, Gao G, Shi L, Lv Y (2017) Kinematic analysis and simulation of a 3-DOF robotic manipulator. In: 3rd IEEE international conference on "computational intelligence and communication technology", 2017. IEEE, pp 1–5
13. Liu Y, Wang D, Sun J, Chang L, Ma C, Ge Y, Gao L (2015) Geometric approach for inverse kinematics analysis of 6-Dof serial robot. In: The 2015 IEEE international conference on information and automation, Lijiang, China. IEEE, pp 852–855
14. Patil A, Kulkarni M, Aswale A (2017) Analysis of the inverse kinematics for 5 DOF robot arm using D-H parameters. In: The 2017 IEEE international conference on real-time computing and robotics. Okinawa, Japan. IEEE, pp 688–693
15. Xu J, Song K, He Y, Dong Z, Yan Y (2018) Inverse kinematics for 6-DOF serial manipulators with offset or reduced wrists via a hierarchical iterative algorithm. IEEE Access 6:52899–52910

Design and Implementation of Intelligent Anti Drunk Driving and Overspeed System

Jiali Song and Shiyu Shang

Abstract Cars have become an indispensable means of transportation in people's life, saving people time and improving efficiency, but traffic accidents caused by drunk and speeding occur frequently. Therefore, it is necessary to design and implement an alarm system for drunk driving detection and overspeed monitoring. When the driver enters the car, he needs to have an alcohol test first, and the car is allowed to ignite after passing the test; When the vehicle is running, the system is based on the positioning and speed measurement function of GPS, and the vehicle speed is monitored in real time through the GPS speed measurement and positioning module. At the same time, the LCD displays the actual vehicle speed monitored by the system and the set safe speed. If it is determined that the vehicle speed exceeds the maximum value set by the system, the buzzer will sound and the warning lamp will flash continuously to remind the driver to slow down. The system can also carry out vehicle positioning through GPS, and then transmit the alarm position through GSM. At this time, supervisors can strengthen the supervision and management of drivers through human intervention to prevent safety accidents.

Keywords Alcohol detection · Positioning module · GPS

1 Introduction

In recent years, with the rapid development of social economy, the continuous improvement of people's living standards and the increase of car ownership year by year, with the continuous improvement of car speed and people's further concern for the living environment, more and more people in the society begin to pay attention to the management of car driving safety and road traffic safety [1]. However, the incidence of traffic accidents all over the world has not been effectively controlled, and the main factor causing the frequent occurrence of traffic accidents is the driver's

J. Song (✉) · S. Shang
Harbin Vocational & Technical College, Harbin, China
e-mail: 178407833@qq.com

behavior of drunk driving [2]. According to statistics, in 2017, the World Health Organization made statistics on safety accidents caused by drunk driving in various countries. The proportion of deaths caused by drunk driving in many western countries exceeded 30%, which means that nearly one third of drivers died from drunk driving [3, 4]. When drivers drive after drinking, under the stimulation of alcohol, they often suffer from fatigue, drowsiness, inattention, slow response and other phenomena. In case of emergency, they have no time to make correct driving judgment, resulting in passive traffic accidents [5]. In addition, due to the dull coordination of eyes, hands and feet, the handling ability of the car has also decreased, resulting in active accidents. Compared with normal driving, the incidence of traffic accidents caused by drunk driving is more than 16 times [6]. It can be seen that drunk driving not only brings great danger to the personal safety of drivers and their families, but also brings certain potential safety hazards to passers-by. It is necessary to take corresponding compulsory measures to limit the occurrence of drunk driving and ensure people's safety [7, 8]. However, China's current measures to restrict drunk driving still remain in the traditional stage, that is, the traffic police set up cards to detect the alcohol of drivers in passing vehicles [9]. With the help of alcohol tester to detect drivers after drinking, although it limits drivers' drunk driving behavior to a certain extent, it can not fundamentally limit the occurrence of drunk driving behavior [10].

In view of the above problems, starting from the aspects of intelligence and initiative, combined with the principle of on-board alcohol detection, this paper puts forward the technical means of multi-dimensional detection of alcohol concentration in the cab and implementing anti drunk driving measures, that is, the multi-dimensional alcohol sensor is used to detect the exhaled alcohol concentration of the driver. When it is judged that the exhaled alcohol concentration of the driver exceeds a certain concentration, the car takes the initiative to take the self-locking behavior, and informs the family through GSM technology, avoid drunk driving and cheating. In addition, the drunk driving prevention system through this means has low cost, high accuracy, and is not limited by vehicle models, so it has a good application prospect. Therefore, this study not only has strong theoretical significance, but also has a good effect on practical application, which plays a certain role in improving road traffic safety and accelerating the construction of a harmonious and stable society.

2 Overall Scheme Design

All satellite positioning systems used today have the function of measuring the speed of the receiving end. The coordinated processing of several satellites ultimately means that the receiver can directly output the speed of the equipment [11]. If the vehicle speed is higher than the set speed, it is regarded as overspeed. At this time, it is necessary to realize the audible and visual alarm in the car, at the same time, the control module locates the position, and then generate the position information of the car alarm through GSM. The alcohol concentration on the driver is monitored through the sensor. If the alcohol residue meets the standard, the car is allowed to start

Fig. 1 System principle
block diagram

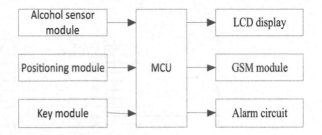

with fire; After passing the alcohol test, start the overspeed test; When the vehicle is running, the system measures the speed through the satellite positioning module based on the speed measurement technology of Beidou, and detects the speed of the vehicle in real time. System safety parameters and are displayed on the display in real time. If the vehicle speed is higher than the maximum setting value, the buzzer will work to show the alarm signal to remind the driver to reduce the vehicle speed; In case of drunk driving or speeding, the system can locate the vehicle through Beidou, and then transmit the alarm information through GSM. The hardware structure of the overall scheme of the specific system is shown in Fig. 1.

3 Hardware Design

MCU circuit design: if stc89c52rc chip is used as the core chip, a crystal oscillator circuit, a reset circuit and a 5 V power supply will be required. The reset circuit is used to clear the internal registers of the single chip microcomputer, especially after opening each device, the system will automatically clear, and then the single chip microcomputer will restart the program. The reset circuit is divided into power reset and button reset [12]. Since it is not necessary to restart the system manually, only a restart is performed. Crystal oscillator circuit is also very important. Without the crystal oscillator circuit, the microcontroller will not work properly [13]. If there is no reset circuit, the system register will not be cleared every time the microcontroller is opened, resulting in confusion when executing the system program. When the system is started, the capacitor is not charged, and there is no voltage on the reset pin of the microcontroller. At this time, the MCU has been turned on, and when the reset pin is low, all registers of the MCU will be initialized. The circuit diagram of single chip microcomputer is shown in Fig. 2.

Alcohol detection circuit design: the basic principle of MQ-3 alcohol sensor can be simply described as a device that converts the detected alcohol concentration into useful electrical signals, and can be measured according to the strength of these electrical signals, so as to provide information about the presence of the gas to be measured [14]. The alcohol detection circuit is shown in Fig. 3.

GSM module circuit design: when the system detects that the exhaled alcohol concentration of the driver exceeds a certain standard value, the anti drunk driving

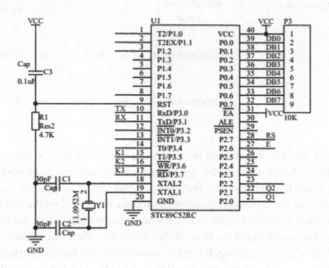

Fig. 2 Schematic diagram of single chip microcomputer circuit

Fig. 3 Circuit diagram of
alcohol detection

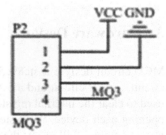

system will send the basic information of the vehicle and the driver's behavior to
the family mobile phone bound to the system through the GSM module, so as to
take corresponding measures. If the family determines that the driver has drinking
behavior and does not determine the system, the vehicle cannot be used normally, if
the vehicle is driven by a substitute driver or a friend and approved by the family, the
vehicle can be used normally. The GSM communication circuit diagram is shown in
Fig. 4.

Positioning module: the GPS module selects the module with gt-u7 as the main
chip. The gt-u7 positioning module is used to determine the longitude and latitude
of the vehicle. The longitude and latitude are displayed on the LCD screen, and the
longitude and latitude can be sent to the family mobile phone bound to the system
through the GSM module, so that the family can understand and determine the
position of the driver driving the vehicle. The specific positioning circuit is shown
in Fig. 5.

Fig. 4 GSM module circuit diagram

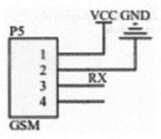

Fig. 5 Positioning module circuit diagram

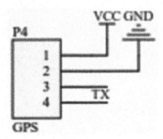

LCD circuit design: LCD12864 is a widely used LCD device. It can not only display images, but also display Chinese characters. The specific circuit of LCD display is shown in Fig. 6.

Alarm circuit design: if the driver in the car drinks too much, the alarm circuit will give an audible and visual alarm through buzzer and LED lamp. When the alcohol sensor of the vehicle detects the corresponding alcohol vapor signal, it transmits the information to the A/D converter and then to the single chip microcomputer for processing. When the alcohol concentration in the car exceeds the programmed threshold, the microcontroller provides the liquid level for the connected connection so that the LED lights up and the buzzer sounds. The sound alarm circuit is shown in Fig. 7.

Fig. 6 LCD display circuit diagram

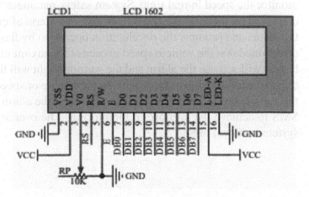

Fig. 7 Circuit diagram of
sound alarm

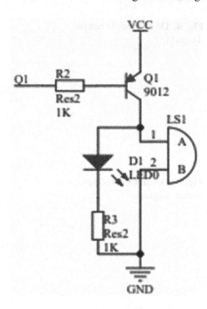

4 Software Design

The design needs to detect whether the driver is drunk driving and speeding. In case of drunk driving and overspeed problems, it will automatically give an alarm and even carry out necessary vehicle braking [15]. Combined with the actual situation of automobile driving, a drunk driving and overspeed alarm control system is designed. When the driver enters the vehicle, first of all, alcohol detection is required to detect whether there is alcohol residue in the driver's mouth and body. If there is no residue, the car is allowed to ignite at this time. If the driver is suspected of drunk driving, the car cannot be started by ignition. After passing the alcohol test, start the overspeed test; When the vehicle is running, the system is based on Beidou's own speed measurement technology and measures the speed through Beidou module to monitor the speed in real time. System safety parameters and real vehicle speed are displayed by the display. The buzzer gives an alarm after the vehicle speed exceeds the value, and prompts the deceleration operation by flashing the alarm light. If it is determined that the vehicle speed exceeds the maximum value set by the driver, the buzzer will activate the alarm and the warning light will flash continuously to remind the driver to slow down. In case of drunk driving or speeding, the system can locate the vehicle through Beidou, and then transmit the alarm information through GSM SMS to achieve the purpose of monitoring. The overall software flow chart of the system is shown in Fig. 8.

Fig. 8 General flow of
system software

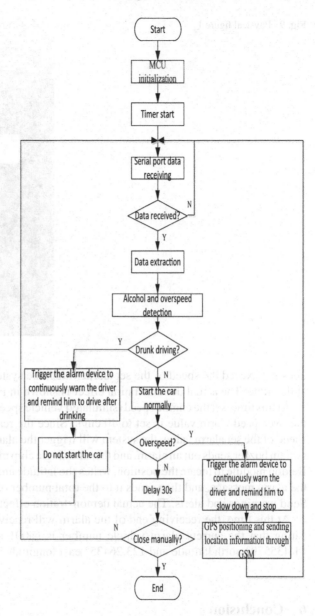

5   Software Design

After hardware welding and software writing, the design will conduct physical tests
on the hardware and software to verify the operation of the whole system. First turn
on the system and set the engine speed (simulated vehicle speed value) to 18 cm/s.
At this time, the overspeed alarm value is set to 40 cm/s. Since the real-time speed

Fig. 9 Physical figure 1

does not exceed the speed of the set alarm value, the system will not give an alarm at this time. The actual demonstration effect is shown in Fig. 9.

At this time, set the engine speed (simulated vehicle speed) to 90 cm/s. At this time, the overspeed alarm value is set to 40 cm/s. Since the real-time speed exceeds the speed of the set alarm value, the system will trigger the alarm at this time. Firstly, the system buzzer sends out an alarm, and then the single chip microcomputer controls the Beidou module to locate the position, writes the latitude and longitude information in the alarm message, and then sends it to the total number of mobile phones received. Send warnings and alerts. The actual demonstration effect is shown in Fig. 10.

At this time, the receiving end of the alarm will receive an alarm message indicating that the current alarm vehicle number is ga001 and the speed position is 41.435579° north latitude and 123.26435° east longitude.

6   Conclusion

The vehicle mounted system based on single chip microcomputer control to prevent vehicle overspeed and drunk driving is adopted, and the modular software and hardware are designed, including several modules such as sound alarm, positioning, speed detection, alcohol detection, GSM transmission and so on. Based on the sensor, speed measurement and positioning module, obtain the alcohol content in the cab, vehicle speed information and vehicle location information, and analyze the drunk driving

Fig. 10 Physical figure 2

and vehicle overspeed. GSM transmission technology is applied to send the location of the problem vehicle to relevant management departments in real time to prevent subsequent accidents. The welding of hardware and the compilation of software are completed. Through experimental verification, it is proved that the judgment of each module is accurate, the simulation analysis and monitoring results are reliable, and the system stability and anti-interference ability are good. Through the steps of overall structure design, selection of hardware components, circuit design of each module, software development and physical production, the system finally obtains a set of systems that meet the functional requirements of the design. After the physical function verification, the physical object meets the design requirements and has a certain use value.

References

1. Tian A (2020) Construction of enterprise management information system. Electron Technol Softw Eng 15:237–238
2. Huang C, Chen Y, Shen J et al (2012) Design and implementation of on-board monitoring and tracking system for drunk driving. Popular Technol 1:16–18
3. Lu Z (2014) Design of embedded drunk driving automatic detection system based on internet of things. Internet of Things Technol 21(7):8–11
4. Li T, Li H, Feng J (2011) Research on vehicle drunk driving automatic detection system based on single chip microcomputer. Sci Times Monthly 2:82–83

5. Sang N, Gan W, Zhou X et al (2018) Design of automobile anti drunk driving control system. J Changzhou Inst Technol 31(03):28–33
6. Wang Q, Yu S (2016) Design of automobile overspeed alarm. Automobile Pract Technol 09:100–102
7. Li Y, Ma H (2016) Design and implementation of anti drunk driving system based on single chip microcomputer. Sci Technol Econ Inner Mongolia 22:82–83
8. Guo D, Li Y (2013) Design of embedded drunk driving automatic detection system based on internet of things. Comput Meas Contr 21(3):549–550
9. Ye X, Zhang Y, Wu J (2017) principle of satellite navigation and positioning system and its application in engineering construction. Electron Test 15:62–63
10. Xu X (2018) Discussion on the development of Beidou satellite navigation system in the field of forestry. Digit Commun World 10:13–14+17
11. Zhang Z (2017) Research on multisensor information fusion and its application. Xi'an University of Electronic Science and technology, Xi'an
12. Gan Z (2018) Design of GPRS electric energy meter remote acquisition module based on sim900a. Electron Tech 47(04):32–36
13. Lv X (2019) Design of alcohol detection system based on MQ-3. Sci Technol Horiz 21:31–32
14. Kuang A (2018) Design of alcohol concentration test system based on MQ-3. Digit. Technol Appl 36(04):161–163
15. Zheng J, Zhao X, Bai F, et al (2016) Design of automatic detection system for drunk driving based on internet of things. Commun World 16:271–272

Modern Energy and Electrical Engineering

Micro-control Type Partition Window Solar Energy Internet of Things Power Supply Device

Weiwei Tao and Liu Guangxu

Abstract There are many ways to transmit power through wireless, especially in the application system of the Internet of Things, when the wireless power supply guarantee of the isolation distance is encountered, the application is more scenarios. On short distances where the isolation is less than a few inches, such as the isolation of glazing, or inductive coupling is commonly used. This article discusses the solution of using inductive coupling on one side of the spacer in an environment with spacers. In a typical inductively coupled wireless power transfer system, an ac magnetic field is generated by the transmit coil, which in turn induces an ac current in the receive coil, just like a typical transformer system, but it is glass that isolates the electromagnetic field. The main difference between glass and wireless power transfer systems is that in wireless power transfer systems, the efficiency may be greatly reduced by separating the transmitter and receiver with a gap composed of air gaps or other non-magnetic materials. Furthermore, the coupling between the transmit and receive coils is typically very weak. Couplings of 0.95–1 are common in transformer systems, but in power transfer systems behind insulating glass, coupling coefficients range from 0.8 to as low as 0.05.

Keywords Solar energy · Internet of things · Micro power supply · Isolator

1 Overviews

At present, the utilization rate of solar cells is very high, and the price is also very cheap. As a new energy battery, it has been very popular in many product development and applications, especially the online solar lamps are cheap and fully controlled. Just buy it, application. However, the current finished solar cell modules are all wired connections, that is, the solar panels and the charging module are connected by wires,

W. Tao
Sichuan Vocational College of Cultural Industries, Chengdu, China

L. Guangxu (✉)
Ziyang Big Data Service Center, Ziyang, China
e-mail: 31409957@qq.com

so that as long as the light reaches the control threshold, the charging mode of the backup battery can be turned on. Even in Chengdu, Sichuan Basin, it is possible to obtain relatively Good solar energy utilization. Although the utilization rate of solar energy is very high, the wire connection is always a defect. Try to use a wireless solution to solve this problem, you can install a wireless kit on the window to connect the solar battery and the charging device for the Internet of Things system. In order to solve the wireless power transmission behind the spacer, it is feasible to conduct experimental research. This article is divided into five parts. The first is to start with the efficiency of wireless charging. Although wireless charging technology has been widely used, it almost adopts the method of close contact and close distance for charging energy transmission will be greatly reduced. To solve the problem of wireless charging of partitions, the wireless charging efficiency is adjusted and tested after experimental application. The second is for the comparison of the schemes. After the test of the charging efficiency of the spacer, the LTC4125/LTC4124 application built with the autoresonant technology has been verified and tested, and the wireless effective charging function of the 20 cm spacer glass can basically be realized. The third is to determine the wireless charging scheme, and finally implement the application goal of wireless charging by optimizing the adjustment scheme. The fourth is the design of the solar wireless charging control scheme. Due to the long-term application in the outdoors, the choice of solar power supply is the most widely used scheme. The scheme design is designed with a PIC16F18313 micro single-chip microcomputer. The circuit is simple and reliable, and the power consumption is very low. The last part is a summary, a brief summary of the scheme designed in this paper, in order to achieve the purpose of the preliminary scheme design of this paper.

2   The Problem of Wireless Charging Efficiency

At present, the wireless transmission technology has matured, mainly by increasing the power to support the transmission power [1]. When the current flows through the transmission coil, a magnetic field is generated, and the strength of the magnetic field is closely related to power, coil matching, intermediate medium and distance. In particular, when alternating current flows through the transmission coil, a changing magnetic field is created around the coil. If another coil is placed in this resulting alternating magnetic field, an alternating current is received in the second coil. The transmission efficiency of wireless charging is related to the magnetic field density generated by the transmission coil and is proportional to the magnitude of the current flowing through the coil. The magnetic field generated from the transmission coil (primary-side magnetic field) has a significant influence on the receiving coil (secondary-side magnetic field) through the above-mentioned magnetic coupling method to form a current for transmission [2]. In a loosely coupled system that is only air, the electromagnetic coupling coefficient is very low, and the high-frequency current will not be transmitted exactly as the preset coil stacks, but will lose energy quickly because the impedance along the coil does not Matching, this will cause the energy to be reflected back to the source, or radiated to the air and lost. For example, adding another medium in the air, even if it is transparent glass, will greatly reduce the energy transmission efficiency. Therefore, under the premise that the solar charging demand can be met by default, the efficiency is improved as much as possible and the transmission power is increased.

3 The Principle of Wireless Charging Switch Conversion

LTC4125 provides a compact circuit to enable wireless charging and discharging of the emission circuit function, for a clearer understanding of wireless charging energy conversion, shown in Fig. 1 as a switch variable change sine to square wave form switch schematic, from SW1, SW2 switch switching, can achieve effective energy conversion, and then through the electromagnetic field reaction conversion, both can get the required energy conversion. Of course, in reality, SW1, SW2 will not use ordinary switch execution, but the use of high-frequency electronic components to replace the way its replacement is built by push–pull circuit [3].

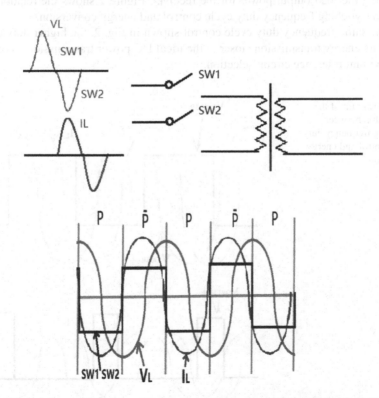

Fig. 1 The switch variable is changed from sine to square wave to form switch schematic

In Fig. 1, the double-open switching mode is used to show the effective segmentation of the sinusoidal waveform. When SW1 is connected, only the upper part of the sinusoidal waveform is connected, this opening allows the coil to propagate as a positive half energy during charging [4]. Similarly, when SW2 is plugged in, only the lower part of the sine wave is plugged in, and the opening of this part enables the coil to propagate negative half-cycle energy during charging. When the cycle is opened, the energy conversion and propagation of sinusoidal full-period wave can be obtained.

Because the sinusoidal positive and negative half energy alternately transforms, forms the transmission uninterrupted energy loop on the wireless charging coil energy dissemination, then can realize the direct current energy transformation to the variable transmission energy, in this way, the DC Energy obtained by the solar energy is transferred by means of electronic switch and coil, so as to achieve the high efficiency of air separation and energy transfer [5]. LTC4125 also adopts this mode for energy transfer and implementation. In practical applications, the device also adjusts the pulse width of the series energy transfer coil waveform by changing the duty cycle of the push–pull full-bridge switching circuit. By increasing the duty cycle, more alternating currents are generated in series in the switching circuit network, thus increasing the load output power for the receiver. Figure 2 shows the relationship between switching frequency duty cycle control and energy conversion.

In the same frequency duty cycle control shown in Fig. 2, the higher duty cycle control its energy transmission closer to the ideal DC power transmission, is one of the important reference circuit selection.

Fig. 2 Diagram of the relationship between switching frequency duty cycle control and energy conversion

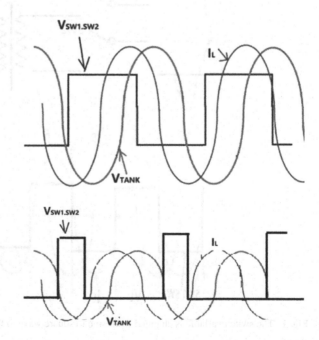

4 Selection of Wireless Charging Solutions

Most of the current wireless charging kits are designed according to the Qi standard, and have clear standards and requirements for the transmission distance and medium [6]. The LTC4125 (transmit)/LTC4124 (receive) kit was chosen to be suitable for miniaturization and wireless charging on windows. This kit is a wireless synchronous buck charger with a solution designed to meet the needs of high reliability applications. Embedded in the LTC4124 is Dynamic Harmonized Control (DHC) tuning technology, which dynamically changes the receiver's resonant frequency to accommodate environmental and load changes [7]. DHC achieves higher power transfer efficiency, allows for smaller receiver size, generates negligible EMI, and even allows for better efficiencies with the addition of a glass medium, for example, to extend the transmission distance equivalently. If the LTC4124 receives more energy than is required to charge the battery, the wireless power manager in the IC can keep the IC's input voltage VCC low by shunting the receiver resonant circuit to ground. During the charging process, the charging efficiency of the linear charger is very efficient, because its input voltage can always be kept just above the battery voltage VBATT to meet the high-voltage state charging, even with small currents. When the shunt circuit is engaged, the receiver resonant frequency will be out of tune with the transmitter frequency and the resonant circuit will therefore receive less energy [8]. Its wireless receiving circuit based on LTC4124 is shown in Fig. 3.

The LTC4125 shown in Fig. 4 is a high performance wireless transmitter circuit with complete protection for wireless charging applications. The optimized power search function in the LTC4125 adjusts the transmit power based on receiver load requirements. The LTC4125 also includes several foreign object detection methods to prevent other objects from receiving unwanted power from the transmitter [9]. When paired with the LTC4124, the LTC4125 full-bridge resonant driver can be converted to a half-bridge driver to take advantage of a finer search step size, allowing the low-power receiver to receive just enough power to charge the battery [10]. When the battery is close to a fully charged state, the LTC4124 enters constant voltage

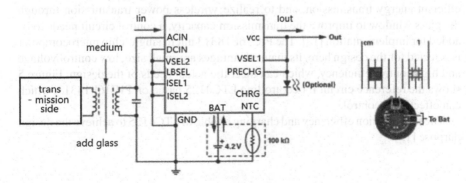

Fig. 3 Wireless receiving circuit based on LTC4124

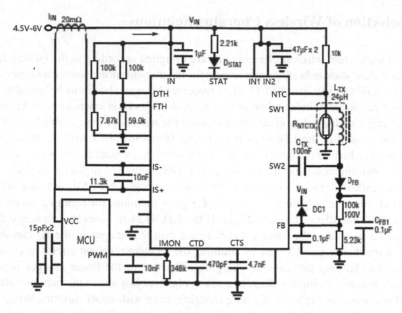

Fig. 4 A high performance wireless transmitter circuit

mode, which reduces the regulated charge current. The LTC4125 will automatically reduce its power delivery level to match the receiver's lower power requirements. This helps reduce power consumption throughout the charge cycle, keeping the LTC4124 charger and battery cooler [11].

5 Solar Wireless Charging Control Scheme

As shown in Fig. 4, in order to enable the wireless charging system to achieve efficient energy transmission, and to realize wireless power transmission through the glass window to improve the transmission capacity, a control circuit needs to be added for implementation [12]. The PIC16F18313 micro single-chip microcomputer is selected for the design here. Its main advantages are small size, low control voltage and high control efficiency, which can meet the actual needs of the design. Figure 5 shows the reference circuit for controlling LTC4125 based on PIC16F18313, which can effectively control.

The transmission efficiency and charging enable of LTC4125 to achieve the design purpose [13].

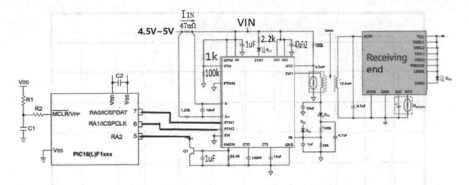

Fig. 5 Reference circuit for controlling LTC4125 based on PIC16F18313

6 Summary

Through the design of this scheme, the function of miniature wireless transmission is achieved, and it can also charge the backup battery in general cloudy days. After testing, the efficiency of the system is reduced by only 13% after the addition of glass partitions under direct sunlight in midsummer at temperatures above 30 °C. Compared with the efficiency, there is only about 20% reduction. It is still basically able to achieve the preset efficiency purpose. The autoresonant architecture of the LTC4125 allows this IC to detect the presence of conductive foreign bodies in a unique way. Conductive foreign matter reduces the effective inductance value in a series lc network. This results in autoresonant drivers increasing the driving frequency of the integrated full bridge circuit.

The graph shown in Fig. 6 compares the frequency of the voltage generated by the transmit coil with and without the presence of a conductive foreign object.

The LTC4125 sets the frequency limit through a resistor divider, reducing the drive pulse width to zero during autoresonant drive beyond this frequency limit. When the LTC4125 detects the presence of a conductive foreign object, it stops delivering power in this way [14].

By using this frequency shift phenomenon to detect the presence of conductive foreign objects, a trade-off between detection sensitivity and component tolerances of the resonant capacitor (c) and transmit coil inductance (l) can be made directly. With a typical initial tolerance of 5% for each l and c value, this frequency limit can be set 10% higher than the natural frequency expected from typical lc values for reasonably sensitive foreign object detection and reliable transmission device circuit design. However, tighter 1% tolerant components can also be used, while the frequency limit is set to only 3% above the expected typical natural frequency to achieve higher detection sensitivity while still maintaining the robust robustness of the design.

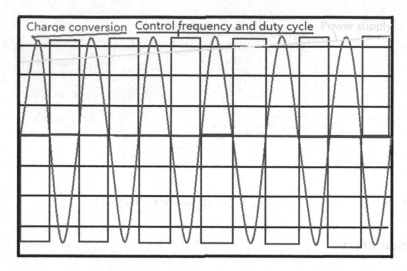

Fig. 6 The relationship between solar radiation, conversion rate and charge rate was measured

References

1. Su Y, Hou X, Dai X (2021) Overview of foreign object detection technology in magnetically coupled wireless power transmission system. Chin J Electr Eng (02)
2. Xue M, Yang Q, Zhang P, Guo J, Li Y, Zhang X (2021) Research status and key issues of wireless power transmission technology application. J Electrotech Technol (08)
3. Jia J, Yan X (2020) Research trends of magnetic coupling resonance wireless power transmission characteristics. J Electrotech Technol (20)
4. Xiao B, Zhou W, Jiang Z, Qi X, Niu X (2020) Analysis of the research status of ubiquitous power internet of things. Power Gener Technol (01)
5. Perol T, Gharbi M, Denolle M (2018) Convolutional neural network for earthquake detection and location. Sci Adv (2)
6. Liu N (2021) Design and implementation of wireless charging system based on QI standard. J Zhengzhou Railw Vocat Tech Coll 33(03)
7. Guo A, Chen F (2020) Deep Q network method for dynamic coordinated control of trunk lines. Inf Technol Netw Secur 39(06)
8. Wei Y, Dong L, Yang W, Tang Y, Zou X (2021) Simulation research on heating and electromagnetic radiation of electromagnetic induction wireless charging. J Radio Wave Sci 36(05)
9. Yao G, Zhao J, Li P (2016) Coil structure design based on QI standard. Sci Technol Innov Appl (21)
10. (2020) ADI ADUM4221 4A isolated half-bridge gate driver solution. World Electron Compon (09)
11. Zhang S, Su C, Zhang K (2021) Design and application of solar wireless charger. Autom Instrum (11)
12. Chen L, Liu Z, Yuan Q, Yang Z (2019) Design of solar wireless charger. Sci Technol Vis (31)
13. Cao K (2018) Design and implementation of a small automatic irrigation system based on single chip microcomputer. Commun Power Technol (03)
14. Zhao Y, Lu B, Zhang M, Liu Y, Liu X, Li W (2019) Research on photovoltaic maximum power tracking technology based on Newton interpolation. J Liaoning Univ Technol (Nat Sci Ed) (03)

Exact Voltage-Drop Correction Factors Applied to Primary Distribution Systems

Nabil Abdel-Karim, Michel Nahas, Georges Semaan, and Fouad Slaoui

Abstract This paper presents an exact method for the correction factors to be used with the voltage drop calculation for distribution primary circuits. The voltage drop in-phase and resistive components are considered separately in this analysis and useful equations are developed to describe the relationship between the correction factor, conductor size and different loading conditions for a given permissible voltage drop level. This novel approach extends beyond lagging loads traditionally considered in the literature and explores leading conditions which are more likely to happen with the large-scale integration of lower emitting variable resources invoking bidirectional power flow among other operational constraints. Circuit limitation in terms of kVA–mile and cable sizing are also investigated and useful equations are developed to demonstrate the fundamental relations between different parameters affecting the voltage drop. Finally, graphical examples will illustrate the relationship between voltage drop, correction factors and other circuit parameters for a square-feed arrangement.

Keywords Correction factor · Distribution systems · Substation area · Voltage drop

List of Symbols

V_s, V_r	Phasor voltage, sending-end and receiving-end, V
I	Phasor (main) current, A
Z	Complex impedance, $R + jX$ in Ω

N. Abdel-Karim (✉) · G. Semaan
ECCE Department, Notre Dame University, Zouk Mosbeh, Lebanon
e-mail: nsabdelkarim@ndu.edu.lb

M. Nahas
EE Department, Australian University, West Mishref, Kuwait

F. Slaoui
EE Department, University of Quebec, Quebec City, QC, Canada

© The Author(s), under exclusive license to Springer Nature Singapore Pte Ltd. 2023
Y. Ma (ed.), *Advanced Theory and Applications of Engineering Systems Under the Framework of Industry 4.0*, https://doi.org/10.1007/978-981-19-9825-6_12

θ	Load power factor angle
%VD	Percent voltage drop between the feed point to most distant transformer
kVA, kV$_{LL}$	Apparent power supplied by the feeder, kV line-to-line voltage
P_m	Feeder active losses, kW
a, d	Feeder length, spacing between laterals, mi
A	Substation service area, mi^2
D	Uniform load density, kVA/mi^2
F_1	Correction factor for resistive voltage-drop
F_2	Correction factor for in-phase voltage-drop

1 Introduction

Distribution utilities are relentlessly looking for ways to increase operation flexibility, system efficiency, reliability, resilience and large-scale integration of higher-level distributed energy resources. The deployment of new technologies tools such as advanced distribution management systems, advanced metering infrastructure, and outage management systems can reshape the future of the utility industry. These tools offer numerous benefits, introduce new business models and also can change the way power networks are designed and operated. The integration of such advanced systems is also critical for utilities to evolve and develop new strategies. Their interconnection and interoperability issues are addressed in IEEE 1547 standard [1]. Despite the major updates in the modern distribution grids, the voltage drop remains an important key parameter to measure the system adequacy regardless of new technologies and tools that are being used or deployed. This parameter must be controlled and kept within an acceptable range set by the utility planners or regulators. For instance, the maximum permissible voltage drop between the substation feed point and the most distant distribution transformer is about 5% for most utilities worldwide. Excessive voltage drop leads to increased system overall losses and reactive flow and causes stability issues. Power quality can also be adversely affected in terms of poor efficiency, regulation errors, and possibly misoperation of intelligent electronic equipment (IEDs), smart sensors and actuators which are nowadays an integral part of the information and communication technology (ICTs) systems used toward the increased digitalization of modern cyber physical grids. Digital communications can help achieve higher resilience levels, better transparency and control capabilities and enhance the distribution system islanding formation when the system is faced with high-impact low-probability extreme events such hurricanes, floods and wildfires. During this transition that is currently taking place worldwide, substations are required to provide many more services like never before with the increased integration of artificial intelligence (AI) and communication protocols based on the IEC61850 dedicated for substations automation and controls [2]. Traditionally, the voltage drop involves several assumptions and approximations. These idealizations introduce a certain amount of error varying in different directions which are usually acceptable for preliminary assessments. Among those, the |ZI| approximation [3] is a well-known approach but do not give accurate results when the substation service area and future load growth become of interest. Many papers dealt with the voltage profile along the primary main [4–6] and feeder reconfiguration based on overall system loss minimization [7, 8]. A voltage profile improvement method involving distributed generation using high-precision measurements (HPMs) connected at the end of the main by considering the load characteristics for radial distribution configurations [9]. Another substation-based method to predict the voltage and current magnitude for remote terminal units (RTUs) was proposed in [10]. Accurate VD-based section loading estimation technique through voltage measurement was discussed in [11] where various loading conditions were simulated and compared. The employment of high-precision phasor measurement units (PMUs) known as synchrophasors for dynamic monitoring are reported in [12, 13]. Strategically located at some selected feed points, these PMUs have the distinct advantage of being able to detect the smallest change in voltage/current waveforms. The huge volume of data generated from the use the of PMUs are then collected, validated, processed and shared for remedial actions. Moreover, the recent high-penetration of distributed energy resources (DERs) also increases the system complexity because of the bidirectional power flow. They create operational issues related to the system voltage stability and the lack of reactive output. Usually, synchronous condensers are installed at specific locations for the necessary Volt/VAR control [14–16]. Another approach based on network classes and load tap changers is developed to identify voltage drop apportionment and improve voltage regulation along with a cost analysis is discussed in [17]. The causes and effects of voltage drops for 11-kV radial systems were also analyzed and reported in [18]. According to the authors the main cause could be poor termination, wires tapering and the use of different conductor materials. Electric vehicles (EVs) market share on the other hand is on the rise and many governmental bodies have pledged to phase out internal combustion engines by 2030. However, excessive voltage drop can occur due to the high demand during the charging process causing stability issues and possibly threatening the grid flexibility. This issue is addressed in [19] where the authors have presented a study on voltage drop compensation by means of reactive injection and adjusted charging schemes.

In many technical references, VD calculation is based on whether omitting the quadrature component or merely by reducing to the resistive component. The proposed approach in this paper introduces two correction factors when the voltage drop is based on either of these approximations. Using these factors allow to predict the exact drop where leading loads are also accounted for. Such loading is a low probability for radial feeders energized from a single feed point but it is likely to occur as today's power delivery systems are experiencing two-way power flow resulting from the integration of inverter-based sources (IBRs) at increasingly higher power level. General VD-based expressions are developed for the substation service area, the feeder kVA-loading and the overall resistive losses. The graphical illustrations show the relationship between the voltage drop, correction factors, substation area, wire size and load density among other geometric factors. The approach presented here is applicable for a square-feed arrangement but it can also be extended to consider different configurations.

2   Exact Voltage Drop Expression

Let a line section of impedance $Z = R + jX$ having a receiving-end current. The functional relationship between voltage and current is:

$$V_s = V_r + ZI \tag{1}$$

Equating magnitudes and arranging give the exact solution for the receiving-end voltage

$$V_r = \left(V_s^2 - (RI\sin\theta + XI\cos\theta)^2\right)^{1/2} - (RI\cos\theta - XI\sin\theta)^{1/2} \tag{2}$$

and the exact voltage drop expression becomes

$$VD = V_s + (RI\cos\theta - XI\sin\theta) - \left(V_s^2 - (RI\sin\theta + XI\cos\theta)^2\right)^{1/2} \tag{3}$$

This expression however requires the knowledge of the primary feeder current which could be circumvented by the introduction of some correction factors that are best understood by referring to the phasor diagram in Fig. 1.

The correction factor F_1 is used to predict the exact voltage drop when estimated by the resistive component and F_2 is used when the quadrature component is neglected. This diagram is referenced in many technical standards and scientific papers and is a useful graphical tools inherent to the voltage drop calculation [20].

Fig. 1 Phasor diagram for voltage-drop calculation

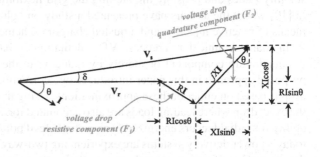

2.1 Correction Factor for Use with the Resistive Voltage Drop Component

When the voltage drop is estimated by the resistive voltage drop component RI, then the correction factor F_1 is introduced as follows:

$$V_s - V_r = VD = RI/F_1 \tag{4}$$

In reference to the phasor diagram and letting $V_{i1} = RI$ and $V_{q1} = XI$, the following nonlinear quadratic equation is obtained:

$$v^2 - (2\gamma_1 \sin\theta)v + \left(2\gamma_1 \cos\theta - 2\gamma_1 F_1^{-1} - F_1^{-2} + 1\right) = 0 \tag{5}$$

where

$$\begin{cases} v = V_{q1}/V_{i1} = X/R = \tan\alpha \\ \gamma_1 = V_r/V_{i1} = 100 - \%\text{VD}/\%\ VD \cdot F_1 \end{cases} \tag{6}$$

The roots of (5) are nothing but to the conductor size often referred to as tanα in power system analysis. Solving this quadratic equation yields to

$$v_{1,2} = \gamma_1 \sin\theta \left(1 \pm \sqrt{1 - c/(\gamma_1 \sin\theta)^2}\right) \tag{7}$$

The c-constant is related to γ_1 and F_1 by $\gamma_1 \cos\theta - 2\gamma_1 F_1^{-1} - F_1^{-2} + 1$.

Knowing that both tanα and the radicand in (7) are both real positive numbers, then the correction factor F_1 is related to the conductor size and the voltage drop by:

$$\begin{cases} \tan\alpha = \gamma_1 \left[\sin\theta \pm \sqrt{\begin{array}{c} \sin^2\theta + 2(\gamma_1 F_1)^{-1}(1 - F_1\cos\theta) \\ + (\gamma_1 F_1)^2(1 + F_1^2) \end{array}}\right] \\ \textit{subject to} \\ \blacksquare \pm \left(2\gamma_1(\cos\theta - F_1^{-1}) - F_1^{-2} + 1\right) \le 0 \\ \blacksquare 2\gamma_1 \cos\theta - \gamma_1^2 \sin^2\theta - F_1^{-1}(2\gamma_1 + F_1^{-1}) + 1 \le 0 \end{cases} \tag{8}$$

The ± in the inequality constraint apply for lagging and leading loads respectively. The variation of F_1 versus the conductor size is given Fig. 2 for different loading levels and 5% voltage drop assumed positive in the direction of the current flow. Obviously, the error using the RI estimation method is conservative for small conductors and high lagging loads and becomes more serious for large conductors irrespective the loading conditions. For instance, if a #2/0 ACSR circuit with 10-feet spacing between conductors is loaded at 0.95 PF lagging to the limiting voltage drop of 5% as determined by the RI method, then the actual voltage drop at this

Fig. 2 Correction factor F_1 as a function of conductor size

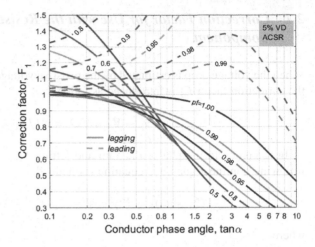

Fig. 3 Correction factor F_1 versus the voltage drop

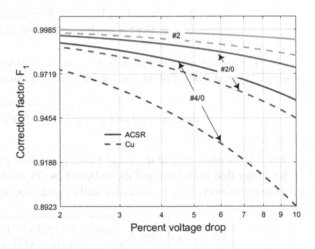

loading is 5/0.72 or 6.89%. In the case of pure resistive loads, then Eq. (8) is could be approximated by

$$\tan\alpha = 14.23\sqrt{\%VD^{-1}(1 - F_1)F_1^{-2}} \qquad (9)$$

Figure 3 shows the variation of the correction factor versus the voltage drop for copper and ACSR conductors of different sizes ordinary used for primary delivery systems.

As expected, the voltage drop and the correction factor vary inversely and the relative position of these curves indicate that ACSR conductors produce better results. The error becomes more important for large wires assuming the same permissible voltage drop and the same spacing between conductors. This is evident because of the high X/R ratio.

2.2 Correction Factor for Use When Neglecting the Quadrature Component

Letting now $V_{i2} = RI\cos\theta - XI\sin\theta$ and $V_{q2} = RI\sin\theta + XI\cos\theta$, then Eq. (1) can be rewritten in terms of the in-phase and quadrature components as follows:

$$V_s = (V_r + V_{i2}) + jV_{q2} \tag{10}$$

The error between the estimated voltage drop and the actual voltage drop is depicted in Fig. 4 which is the horizontal distance between V_{i2} and the circumference of the circle that has a radius of Vs.

A further simplification is made to the horizontal component regarding to the term $-XI\sin\theta$. For lagging loads, both θ and $\sin\theta$ are negative, and thus the voltage drop due to its reactive component is positive and the in-phase voltage drop component is now written as $RI\cos\theta + XI\sin\theta$. By restricting the voltage drop to the in-phase component, then the correction factor F_2 as follows:

$$V_s - V_r = VD = V_{i2}/F_2 \tag{11}$$

Taking similar steps as previously results in the following quadratic expression:

$$v_2^2 - 2v_2(F_2^{-1} - 1) - F_2^{-1} + 1 = 0 \tag{12}$$

where

Fig. 4 Error illustration when using the RI method

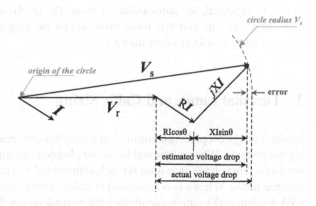

Fig. 5 Correction factor
F_2 as a function of conductor
size

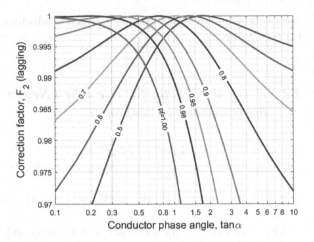

where $\begin{cases} v_2 = V_{q2}/V_{i2} = \sin\theta + \tan\alpha\,\cos\theta/\cos\theta - \tan\alpha\,\sin\theta \\ \gamma_2 = V_r/V_{i2} = 100 - \%VD/\%VD \cdot F_2 \end{cases}$ (13)

Solving for $\tan\alpha$ in terms of other circuit parameters

$$\tan\alpha = (k_2 - \tan\theta)/(1 + k_2\,\tan\theta) \qquad (14)$$

where $k_2^2 = F_2^{-2} - 1 + 2(100 - \%VD)(F_2^{-1} - 1)(\%VD \cdot F_2)^{-1}$

The error introduced using this method is not appreciable since the correction factor is close to unity in most cases as indicated in Fig. 5. For example, if a 500-MCM ACSR circuit is loaded at 0.98 PF to the limiting voltage drop of 5%, then the actual voltage drop at this loading is 5/0.9857 or 5.07%. Within the broad scope of planning, this error is not serious and is in the conservative direction.

Figure 6 is a comparative illustration between F_1 and F_2 for both loading conditions. As expected, the approximation using the in-phase voltage drop component gives better results and it is more pronounced for lagging loads as F_2 varies in a close-to-unity narrower range than F_1.

3 Thermal Limits and Cable Sizing

Beside voltage drop, thermal limitation is another concern when operating the distribution systems. When additional loads are planned, the circuits are usually extended until a certain voltage drop limit is reached provided that the circuit operates below its thermal limits. When a new feed point is added to an existing distribution system, the kVA amount and location can disrupt the normal power flow and distort the voltage profile at the same time. The kVA-mi loading can be found by:

Fig. 6 Correction factor F_2 versus F_1

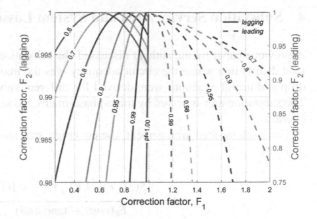

$$kVA \cdot d = \frac{kV_{LL}^2}{0.1r\,F_2(\cos\theta - \tan\alpha\sin\theta)}\left(\frac{V_{i,2}}{V_s} \times 100\right)$$

(15)

This equation is useful in comparing the economics of various plans of distribution systems reinforcement. Note that the term between brackets is the approximated in-phase voltage drop. The kVA flow is not greatly affected by the power factor for distribution circuits generally characterized by a $\tan\alpha$ less than 1.2 as illustrated in Fig. 7 drawn for different operating voltage levels. Assuming that the conductor size is proportional to its Ampere loading, the sizing error of the conductor cross sectional area is about $100(1 - F_2)$. Since F_2 is less than unity for all power factors, wires are then undersized when the voltage drop is approximated by the in-phase component V_{i2}.

Fig. 7 Substation KVA-miles loading at 5% VD

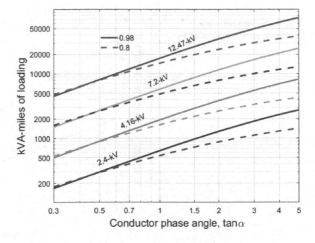

4 Substation Service Area and System Losses

A previous paper by the authors discussed the substation service area for uniform load distribution where the circuit adequacy was the maximum admissible voltage drop and useful equations were derived for different substation area patterns [21]. The substation area is served by three-phase mains and single-phase multigrounded laterals.

For a square-feed arrangement, the coverage area and the associated losses are given by:

$$A = \frac{15(D \cdot r)^{-1}kV_{LL}^2\left(\frac{V_{i,2}}{V_s} \times 100\right)}{F_2(\cos\theta - \tan\alpha\sin\theta)} \tag{16}$$

$$P_m = \frac{480r^{-1}kV_{LL}^2\left(\frac{V_{i,2}}{V_s} \times 100\right)^2}{aF_2^2(\cos\theta - \tan\alpha\sin\theta)^2} \tag{17}$$

These equations, reveal that the correction factor F2 affects both the substation area and the I^2R losses as illustrated by Fig. 8 drawn for a 4.16 kV$_{LL}$ system and 5% voltage drop.

As expected, higher load densities result in increased active losses and smaller area coverage for a given correction factor. Placing a dollar value on these losses would probably represent a more useful approach for economic studies. These studies also include energy and demand charges and the cost of installed conductors. However, this may not be conclusive as the substation cost and other technical factors must be considered in true economic analysis which beyond the scope of this paper (Fig. 9).

Fig. 8 Substation area coverage as a function of F$_2$

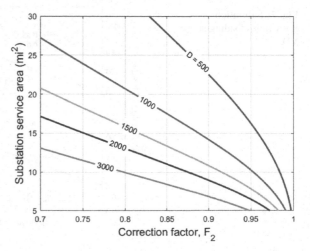

Fig. 9 Total resistive feeder loss

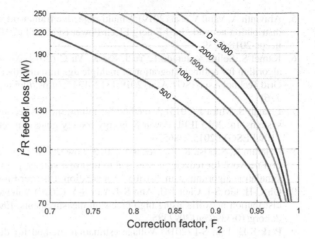

5 Conclusion

Voltage drop is a serious concern for utility planners. Numerous approaches were developed where the level of accuracy can vary greatly. This parameter is considered as an adequacy measure and is generally approximated leading to some significant calculation errors. In this paper, equations for voltage drop correction factors were developed to illustrate the fundamental relationship between wire sizes, loading conditions and their effect on the approximated voltage drop. The concept of correction factors for voltage-drop limited feeders is a useful approach for distribution planners who are constantly faced with the problem of growing demand. Substation area expansion, kVA-miles loading and I^2R loss equations were also derived for the square-feed arrangement as a function of the correction factor, voltage drop and other circuit parameters. Several graphs are plotted to illustrate the relationship between several parameters affecting the voltage drop for primary distribution systems.

References

1. IEEE standard for interconnection and interoperability for distributed energy resources with associated electric power systems interfaces, IEEE standards coordinating committee IEEE Std 1547-2018
2. Communication networks and systems for power utility automation, International Electrotechnical Committee IEC 61850:2022 SER series
3. IEEE recommended practice for electric power distribution for industrial plants, IEEE standards coordinating committee IEEE Std-141-1993
4. Tolba M, Tulsky V (2021) Integration of DGs optimally to enhance the voltage profile and stability index of distribution grid. In: IEEE 3rd International youth conference on radio electronics, electrical and power engineering, Russia, March 2021

5. Aravinth A, Vatul V et al (2019) A multi-objective framework to improve voltage stability in a distribution network. Int J Emerg Electr Power Syst 20(1):20180239. https://doi.org/10.1515/ijeeps-2018-0239
6. Kamel S, Selim A, Jurado F, Yu J, Xie K, Yu C (2019) Multi-objective whale optimization algorithm for optimal integration of multiple dgs into distribution systems. IEEE Innov Smart Grid Technol Asia (ISGT Asia) 2019:1312–1317. https://doi.org/10.1109/ISGT-Asia.2019.8881761
7. Elsaiah S, Mitra J (2015) A method for minimum loss reconfiguration of radial distribution systems. In: 2015 IEEE Power & energy society general meeting, pp 1–5. https://doi.org/10.1109/PESGM.2015.7286576
8. Rajaram R, Sathisu K, Rajasekar N (2015) Power system reconfiguration in a radial distribution network for reducing losses and to improve voltage profile using modified plant growth simulation algorithm with distributed generation. Elsevier Energy Rep 1:116–122
9. Oh C-H, Go S-I, Choi J-H, Ahn S-J, Yun S-Y (2020) Voltage estimation method for power distribution networks using high-precision measurements. Energies 9:2385 [Online]. https://doi.org/10.3390/en13092385
10. Park S-H, Lim S-I (2018) Voltage estimation method for distribution line with irregularly dispersed load. Trans Korean Inst Electr Eng 67:491–497
11. Park J-Y, Jeon C-W, Lim S-I (2016) Accurate section loading estimation method based on voltage measurement error compensation in distribution systems. J Korean Inst Illum Electr Install Eng 30:43–48
12. Muscas C, Sulis S, Angioni A, Ponci F, Monti A (2014) Impact of different uncertainty sources on a three-phase state estimator for distribution networks. IEEE Trans Instrum Meas 63(9):2200–2209. https://doi.org/10.1109/TIM.2014.2308352
13. Muscas C, Pau M, Pegoraro PA, Sulis S (2016) Uncertainty of voltage profile in PMU-based distribution system state estimation. IEEE Trans Instrum Meas 65(5):988–998. https://doi.org/10.1109/TIM.2015.2494619
14. Dong J, Xue Y, Olama M, Kuruganti T, Nutaro J, Winstead C (2018) Distribution voltage control: current status and future trends. In: 2018 9th IEEE International symposium on power electronics for distributed generation systems (PEDG), pp 1–7. https://doi.org/10.1109/PEDG.2018.8447628
15. Ikeda S, Ohmori H (2018) Evaluation for maximum hosting capacity of distributed generation considering active network management. Int J Electr Electron Eng Telecommun 7(3):96–102. https://doi.org/10.18178/ijeetc.7.3.96-102
16. Tungadio DH, Sun Y (2020) Active power management of islanded interconnected distributed generation. Int J Electr Electron Eng Telecommun 9(3):177–182. https://doi.org/10.18178/ijeetc.9.3.177-182
17. Carter-Brown CG, Gaunt CT (2006) Voltage management by the apportionment of total voltage drop in the planning and operation of combined medium and low voltage distribution systems. SAIEE Africa Res J 97(1):66–73. https://doi.org/10.23919/SAIEE.2006.9488027
18. Nunoo S, Attachie JC, Duah FN (2012) An investigation into the causes and effects of voltage drops on an 11 kV feeder. Can J Electr Electron Eng 3(1):40–47
19. Mitsukuri Y, Hara R, Kita H, Kamiya E, Taki S, Hiraiwa N, Kogure E (2014) A study on compensating voltage drop in distribution systems due to nighttime simultaneous charging of electric vehicles utilizing charging power adjustment and reactive power injection. Electr Eng Jpn 189(4)
20. IEEE recommended practice for electric power systems in commercial buildings, IEEE standards coordinating committee IEEE Std-241-1990
21. Abdel-Karim N, Georges S, Slaoui F (2019) Planning of capacity additions for voltage-drop limited primary distribution systems. Int Rev Electr Eng IREE 14(4)

The Action Analysis of Losing Voltage in Adjacent Substations Caused by Mistaken Bus Grounding

Bin Li, Adili Balati, Qidi Chen, Wei Cao, Jinlong Tan, and Mengfan Zhang

Abstract Power system fault has brought great disturbance to the stable operation of power system. The in-depth search and analysis of fault causes is the basic requirement to improve the reliability of power system. With an example of relay protection action of a power plant due to the electrification of the operator is mismatched with the busbar grounding switch. In this paper, the protection device and automatic bus transfer equipment action in this action case are analyzed from the logic judgment of the protection action and the calculation of the fault information, and the reasons for the loss of pressure in the substation are discussed. It provides reference methods for analysis and treatment of similar accidents to ensuring safe and reliable operation of power grid.

Keywords Operation · Substation voltage loss · Relay protection · Fault analysis · Setting calculation

1 Introduction

Under the current situation, the stable operation of the power system is an important prerequisite to ensure the people's good life. In terms of power system operation and maintenance, the operators often need to operate the disconnector and grounding knife switch frequently. Although the power grid system has formulated a series of rules and regulations to standardize the operation process, it is also common for errors in the operation process to lead to power grid accidents [1]. Relay protection

B. Li (✉) · Q. Chen · J. Tan
Electric Power Research Institute of State Grid Xinjiang Electric Power Co., Ltd.,
Urumqi 830000, Xinjiang, China
e-mail: 1071311634@qq.com

A. Balati · W. Cao
Dispatch Control Center of State Grid Xinjiang Electric Power Co., Ltd., Urumqi 830000,
Xinjiang, China

M. Zhang
Ultrahigh Voltage Subsidiary of State Grid Xinjiang Electric Power Co., Ltd., Urumqi 830000,
Xinjiang, China

© The Author(s), under exclusive license to Springer Nature Singapore Pte Ltd. 2023   159
Y. Ma (ed.), *Advanced Theory and Applications of Engineering Systems Under
the Framework of Industry 4.0*, https://doi.org/10.1007/978-981-19-9825-6_13

is the first line of defense of the power grid, which undertakes the important task of quickly isolating faults for the first time, that is also very important for the operation and maintenance of relay protection [2–5]. Combined with an accident of voltage loss in adjacent plants and stations caused by the operator mistakenly closing the bus grounding knife switch, this paper goes deep into the problems existing in the field, and a reference for the analysis of such faults in the future is provided, some suggestions for the operation and maintenance of relay protection in power system are also put forward.

2   Accident Overview

The operators of the A 110 kV power plant in a certain area incorrectly closed the I side ground knife switch of the 6 kV bus tie switch of busbar I and busbar II, then the 6 kV bus three-phase short circuit fault occurred in the station. The No. 1 main transformer of the A power plant backup protection started, but without fast action. The line protection in the side of B station of the 110 kV line 2 was on, but no action. The distance three-section line protection in the side of C station of the 110 kV line 1 acted to trip, while locking the heavy closure. The backup automatic switch of the 110 kV incoming line in the 110 kV B station action is unsuccessful, causing the 110 kV B station, E station and other substations loss-of-voltage. Some of the power grid in the region are shown in the Fig. 1.

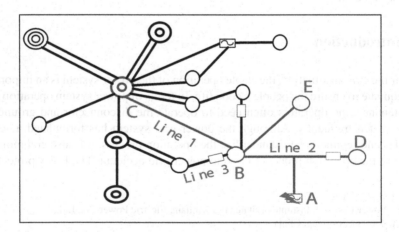

Fig. 1 The system diagram of some of the power grid in the region

3   Operation Method and Grounding Method Before the Accident

The 110 kV system operation mode: no bus; 110 kV line 2 T connecting branch with No. 1 main transformer.

The 6 kV system operation mode: three-stage single bus; the No. 1 main transformer, the No. 1 generator and the excitation transformer were running at the 6 kV I bus; the 6 kV low transformer, the No. 1 low plant transformer, the No. 2 low plant transformer, the No. 1 fan, the No. 2 fan were running at the in 6 kV II bus; the low voltage side of 10 kV auxiliary transformer powered by 10 kV line and capacitor were running at the 6 kV III bus; the bus coupler of the 6 kV busbar I and busbar II was in the quantile, the 6 kV busbar II and busbar III run separately; the bus coupler of the 6 kV busbar II and busbar III was in closing position, the 6 kV busbar I and busbar II run at parallel operation. The main wiring diagram of the A power plant is shown in Fig. 2.

The neutral point grounding method: the 110 kV side neutral point of the No. 1 main transformer was directly grounded.

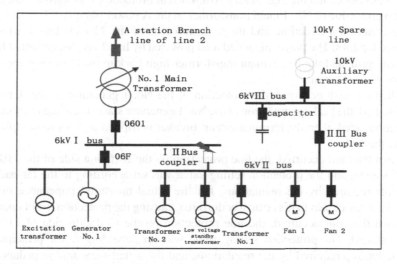

Fig. 2 The main wiring diagram of the A power plant

4 The Process and Protective Action of the Accident

4.1 Primary Equipment Failure

The operators on duty of the A 110 kV power plant in a certain area incorrectly closed the I side ground knife switch electrically of the 6 kV bus tie switch of busbar I and busbar II, causing the 6 kV I bus powered by the No. 1 step-up transformer three-phase short circuit fault.

4.2 Relay Protection Action Situation

The 6 kV bus three-phase short circuit fault occurred in the A power plant. Since the 6 kV bus is not equipped with a bus differential protection device, it cannot act instantaneously to isolate the fault.

At the time of the failure, the No. 1 main transformer protection of the A power plant was started, and the high-reserve overcurrent protection was started. Since the backup protection of No. 1 main transformer in the A power plant is 3.4 s according to the connection direction, and the C station side of the 110 kV line 1 acted in advance for 2.8 s. The power plant and load powered by line 1 are disconnected from the main grid, and the No. 1 main transformer high backup overcurrent protection returns without action.

When the fault occurred, the protection of the No. 1 generator of the A power plant started, the backup protection of the No. 1 generator started, and the overcurrent protection acted, then the terminal circuit breaker is tripped at 5.5 s to completely isolate the fault.

When the fault occurred, the line protection on the B station side of the 110 kV line 2 started, and the protection setting value was set according to the far-backup principle required by the regulations, but the actual measured impedance of the protection device was in it's critical action area, causing the protection to not operate.

When the fault occurred, the line protection on the C station side of 110 kV line 1 started. The protection setting value was set according to the principle of remote backup required by the regulations, and the actual measured impedance of the protection device was in it's action zone of the protection device. The grounding distance section III and phase-to-phase distance section III protection acted to trip, the main grid isolated the fault, and the recloser is blocked (The setting value was set according to the blocking and reclosing of stage III and above in the protection scheme).

4.3 Adaptive Standby Operation and Action

The 110 kV station B was equipped with a set of 110 kV incoming line, bus tie adaptive standby device. The 110 kV line 1 and the bus tie 1150 circuit breaker were in operation. The 110 kV line 3 was in standby state, and the 110 kV incoming line was input to prepare self-switching function.

2.8 s after the fault occurred, the backup automatic switch of the 110 kV incoming line acted in the 110 kV station B. After 4.0 s, the outlet tripped the circuit breaker on the B side of the 110 kV line 1, and the parallel trip outlet tripped the circuit breaker on the B side of the 110 kV line 2. Due to the fact that the check II bus coupler tripping switch position was invested in the setting value of the backup automatic switch of the 110 kV incoming line, but the actual wiring was inconsistent with the on-site wiring, resulting in a misjudgment that the jumper of the tie jumper II bus switch did not return during the operation of the backup switch, which eventually caused the failed preparation operation, and the circuit breaker on the B station side of the 110 kV line 3 was closed before the export, resulting in a loss of voltage at the 110 kV station B.

5 Cause Analysis of Tripping

5.1 110 kV Station B Relay Protection Action Analysis

Through the calculation of the fault quantity information in the oscillogram of the 110 kV line 2 protection device, as shown in Fig. 3, the B station side protection device fault measurement impedance of the 110 kV line 2 was $13.375\angle86°\Omega$ (secondary value, the protection CT ratio was 600/5). The action characteristic was the characteristic of the upper flat quadrilateral, and the action characteristic of the protection device is shown inside the blue edge in the fault analysis diagram in Fig. 4; The setting value of line device distance III section was set according to the remote backup requirements [6]. To protection the low-voltage side fault of the No. 1 main transformer in the A power plant, the setting value was set to 1.2 times sensitivity as $15.4\angle57°\Omega$ (secondary value), and the time is 1.9 s. The green dots are shown in the fault analysis diagram in Fig. 4. According to the action characteristics of the protection device, the setting value converted to an impedance angle of 86° was $12.95\angle86°\Omega$ (secondary value), the protection setting value was in the critical non-action zone of the device action zone, resulting in the protection no action.

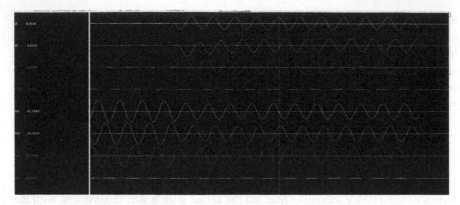

Fig. 3 The B station side of the 110 kV line 2 protection device fault recorder

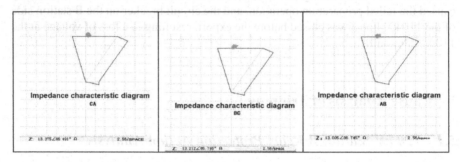

Fig. 4 Fault analysis diagram of the B station side of the 110 kV line 2

5.2 Analysis of the Relay Protection Action of the 110 kV Line 1 of the 220 kV Station C

Calculated according to the fault quantity information in the oscillogram of the 110 kV line 1 protection device of the 220 kV station C, as shown in Fig. 5, the fault current was 1.15A (secondary value, CT ratio was 600/1), the fault voltage was 102.36 V (secondary value, PT ratio was 110/0.1), then it is calculated that the measured fault impedance of the device was $88.58\angle 85°\Omega$ (secondary value), and the action characteristic was the characteristic of the upper hypotenuse quadrilateral. The action characteristics of the protection device are shown inside the blue edge in the fault analysis diagram in Fig. 6. The setting value of the 110 kV line 1 distance section III was in accordance with the requirements of the regulations for remote backup. To protection the low-voltage side fault of the No. 2 main transformer in the B power plant, the setting value was set to 1.2 times sensitivity as $100.4\angle 71°\Omega$ (secondary value), and the time is 2.8 s, as shown by the green dot in the fault analysis diagram in Fig. 6. According to the action characteristics of the protection device, the setting value converted to an impedance angle of 85°was $103.5\angle 85°\Omega$ (secondary value). The protection setting value was in the reliable action area of the device, and at the same time, the B station side of line 2 was in the critical action area but no action, resulting in the protection action of the C station side of the upper 110 kV line 1 delay reaching the time to trip.

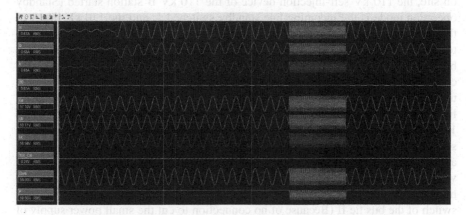

Fig. 5 Fault recording diagram of the protection device on the station A side of line 1

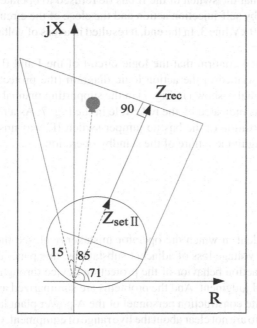

Fig. 6 Fault analysis diagram on the station A side of line 1

5.3   Analysis of the 110 kV Self-switching Action of the 110 kV B Station

From the analysis of the action information of the standby automatic switching device on site, the 110 kV self-injection device of the 110 kV B station started (Standby operation mode is the incoming line standby switched 1, and the 110 kV line 1 of the second power supply was used as the main power supply, which is connected to the 110 kV II bus. The 110 kV line 3 of the first power supply was used as a backup power supply, and the circuit breaker was in a hot standby state, which was connected to the 110 kV I bus). After 4001 ms, the backup switching device acted the chasing jump outlet to trip the circuit breaker of the B station side of 110 kV line 1, and the joint jumping outlet to trip the circuit breaker of the B station side of 110 kV line 2 (small power supply). After the circuit breaker was tripped, the standby self-switching device has received the return input signal of the circuit breaker trip, and this process conforms to the action logic of the protection device. Since the fixed value of the II bus tie jumper switch position was also put into the device setting, but the device has not detected the return input signal of the trip position of the trip switch of the bus tie II (Because of no connection to cut the small power supply of the II bus, the relevant secondary circuit was not connected on site). Therefore, the device misjudged that the switch of the II bus tie refused to operate, resulting in the failure of the standby self-injection action and the close of the circuit breaker on the B station side of 110 kV line 3. In the end, it resulted in a loss of voltage at the 110 kV station B.

Further check and confirm that the logic circuit of the I and II bus tie jumpers was incorrectly described in the action logic diagram (the protection action logic diagram of the manual is shown in Fig. 7) in the supporting manual of the automatic switching device, as indicated by the red circle in the Fig. 7. As a result, the control word "check the position of the bus tie jumper switch II" was mistakenly put into the setting, resulting in the failure of the standby operation.

6   Conclusion

Aiming at an accident in which the operator mistakenly closed the bus grounding switch resulting in voltage loss of adjacent substations, this paper demonstrates the correctness of the action behavior of the protection device through theoretical data analysis and logical judgment. And the problems are summarized as follows: ① The operators and on-site construction personnel of the A power plant lack awareness of safety risks, and who are not clear about the live range of equipment. On-site operators mistakenly close the grounding switch, resulting in a three-phase short-circuit fault on the 6 kV bus, which almost leads to a serious accident of personnel entering the live interval by mistake. ② The A power plant protection configuration is not perfect, the setting is unreasonable. Failure to comply with the requirement "The protection

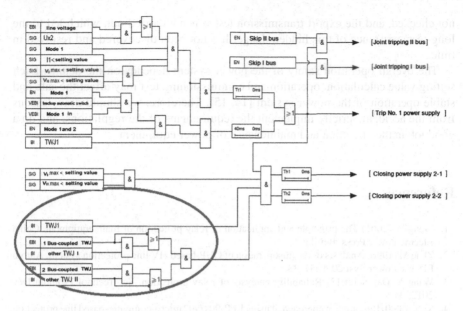

Fig. 7 The logic diagram of the protection action of the self-switching device

of the busbar is realized by the backup protection of generators and transformers" of Article 4.8.3 of the "Technical Regulations for Relay Protection and Safety Automatic Devices". The No. 1 main transformer backup protection device has not set the protection value of the grid direction, and has not considered the configuration of the 6 kV busbar protection device, which actually caused the 6 kV busbar to run without protection, and will inevitably leap to the grid side when a fault occurs [7–9]. ③ The 110 kV line 2 protection setting of 110 kV station B was not optimized and adjusted in combination with the operating characteristics of the device. The setting value of the distance section III protection at the side of the B station 110 kV line 2 is set according to 1.2 times the sensitivity, which does not violate the stipulated in the regulations "The sensitivity can not less than 1.2 times", but has not been further optimized in combination with the operating characteristics of the distance section III protection of this line protection device, causing the actual measured impedance of the protection device to fall on the boundary of the action characteristics of the remote backup protection device, and the sensitivity is insufficient [10–13]. ④ The 110 kV equipment self-switching manual logic diagram error of the 110 kV station B was inconsistent with the actual device principle. The manual of relay protection and safety automatic device is the core technical data and key basis for the setting calculation. It's principle function and setting value description must be consistent with the actual device. In this accident, the error of the logic diagram of the automatic switching manual is an important reason for the setting value error. ⑤ After the 110 kV self-switching value of 110 kV station B was issued to the site, the functional test items of the standby self-switching device were incomplete, the actual wiring was not checked, and the export transmission test was not carried out, which led to the long-term existence of this hidden danger has not been discovered and resolved in time.

The overall operation safety of the power system depends on the details. Each setting value calculation, operation and commissioning test may affect the safe and stable operation of the power system [14, 15]. Therefore, we should draw lessons from the accident, strictly implement the requirements of the regulations, and do a good job in the operation and maintenance of power equipment.

References

1. Wang W (2001) The principle and application of relay protection of main equipment. China Electric Power Press, Beijing
2. Zhao M (1996) Analysis of the misoperation of CUP-based HV line protection system. Autom Electric Power Syst 20(4):31–33
3. Wang Y, Dai X (2012) Reliability analysis of relay protection. Sci Technol Promote Dev 2012(s1)
4. Ye Y (2000) Research on approach of using LFP-901/902 micro-computer-based line protection on series capacitor compensated lines. Autom Electric Power Syst 24(10):25–28
5. Ye Y (2004) Research on the criterion of transformer differential protection. J Anhui Electr Eng Prof Tech College (1):2–6
6. Ye Y (2006) The research on the setting and disposition scheme of the transformer medium & low voltage back-up protection. Power Syst Technol 30(16):25–28
7. Jiao Y (2010) Analysis of differential protection operation caused by no-load transformer. Power Syst Clean Energy 18(2):18–20
8. Li J (2008) Analysis on several problems about transformer ratio-differential relay protection. Jiangsu Electr Appar 22(11):25–30
9. Xu J (2001) A microcomputer system on a transformer differential protection based on fuzzy set and scalar product breaking theory. Relay 29(9):8–11
10. Li D (2004) A regulating method of differential protection with micro-processor for transformer. Installation 8(1):12–14
11. Yu L, Xu Q, Li J (2002) Features analysis on WGYC-1A type microcomputer overvoltage & overexcitation protection. Northeast Electr Power Technol (7):48–50
12. Li X (1999) Over-excitation study of generator-transformer unit. Shanxi Electr Power 89(6):23–59
13. Yang Y (2017) Analysis on reliability improvement of relay protection in thermal power plant. China Equip Eng 2017(13)
14. Li J (2017) Research on improving relay protection reliability of 10kV power supply system. Smart City 2017(07)
15. Song J (2016) Discussion on influencing factors of relay protection reliability. Coal 2016(11)

LSTM Short-Term Load-Prediction Model for IES Including Electricity Price and Attention Mechanism

Yuemao Sheng, Rongqiang Feng, and Hai Lu

Abstract In order to improve the accuracy of short-term load prediction in an integrated energy system, the data of a comprehensive energy demonstration area is analyzed. In the case of similar time-sharing gas prices, the relationship between electricity price and load is verified by using the electricity price and load curve, and the electricity price and gas curve. The influence of electricity price on load prediction is considered. A short term Load Prediction model based on attention-LSTM (Attention Long short term Memory) network is proposed. First, the eigenvectors considering price fluctuations are trained from the input layer to the hidden layer of the LSTM model. The trained eigenvectors are then used as input to the Attention layer to generate the weight vectors. Finally, the eigenvectors and weight vectors are combined to get new vectors, and the predicted values are obtained by training the fully connected layers. The experimental results show that the proposed method has higher load prediction accuracy.

Keywords Integrated energy system · Load prediction · Mutual information · LSTM · Attention

1 Introduction

For integrated energy systems (**IES**) with electricity, gas and heat loads, energy prices can lead consumers to purchase energy [1]. Load Prediction is an important link in the planning and design of integrated energy system and the optimal dispatch of energy [2]. With the rapid changes in the international energy market and the

Y. Sheng (✉) · R. Feng
System R&D Department, NARI Technology Co. Ltd., Nanjing, China
e-mail: yueguixiademao@163.com

R. Feng
e-mail: frqkycg@163.com

H. Lu
Yunnan Electric Power Research Institute, Kunming, China
e-mail: 924298187@qq.com

© The Author(s), under exclusive license to Springer Nature Singapore Pte Ltd. 2023   169
Y. Ma (ed.), *Advanced Theory and Applications of Engineering Systems Under the Framework of Industry 4.0*, https://doi.org/10.1007/978-981-19-9825-6_14

global economic downturn, changes in energy prices have become an important factor affecting load fluctuation, which increases the difficulty of Load Prediction. Therefore, short-term Load Prediction under real-time electricity price and time-sharing price is of great significance. On the other hand, the scale of new energy market is growing, the number of devices is increasing sharply, the level of intelligence is increasing continuously, and the data quantity and accuracy of load collection data are rising sharply, which provides a data base for short-term Load Prediction. Therefore, in order to optimize the dispatch of comprehensive energy, improving the accuracy of short-term Load Prediction has become an urgent problem to be solved.

In recent years, artificial intelligence methods have made great progress in the field of Load Prediction, in which deep learning is often used because of its outstanding ability to deal with non-linear mappings. Reference [3] Based on convolution neural network, a convolution neural network supporting machine vector regression model is proposed. User cluster grouping is used for load prediction, and the final monthly load is obtained by summing up the results. Document [4]. This paper presents a monthly Load Prediction method based on STL decomposition, which decomposes the load series into trend, season and random components. The common problem with the above methods is that the time dependence of time series data is not considered and there is no time characteristic in Load Prediction. Compared with the above neural networks, the long short term memory (LSTM) takes into account the time factor, has the memory ability, and can effectively learn the regular information in the historical data. Reference [5] uses LSTM model to predict user load prediction points, and then uses heuristic interval prediction algorithm to achieve the optimal solution of global prediction. Although LSTM solves the time characteristic problem of time series data, it treats the eigenvectors of input important data and common data equally in the training process, which has certain influence on the accuracy of Load Prediction. In addition, the above literature mostly predicts single load, failing to take into account the coupling characteristics between different loads in the integrated energy system, the paper [6] builds a short-term multi-load prediction model for regional integrated energy system based on improved long-term and short-term memory neural network. The impact of changes in energy prices on Load Prediction in the energy market environment is ignored in the above literature. The current Attention mechanism has a good application effect in the deep learning of natural language processing [7], which can give attention weights to different input features in the training of neural networks.

In summary, the Attention mechanism is introduced into the Load Prediction calculation, and an Attention-LSTM short-term Load Prediction model is proposed to improve the accuracy of Load Prediction, considering the real-time electricity price and time-sharing price factors, historical time series data, and input feature weights, In the application scenario of integrated energy, it is beneficial to optimize multi-energy scheduling, reduce energy consumption, and improve the economic benefits of integrated energy.

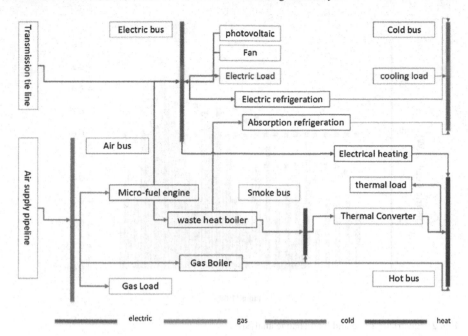

Fig. 1 Integrated energy structure

2   IES and Its Demand Response Model

Shown in the following Fig. 1, IES is connected to power grid and gas network to supply power to the park through distributed generation combined with a variety of energy conversion devices and achieve economic dispatch [8]. The power, gas, cold and heat systems in the park are based on the energy bus model [9].

2.1   Energy Conversion Equipment Model

Combined cooling heating and power: The combined cooling heating and power (CCHP) system generates electricity by burning natural gas in a micro gas turbine and discharging high-temperature gas for heat recovery in a heat recovery boiler and absorption refrigerator [10].

3   Price Guidance Analysis

Real-time electricity price guides the region to purchase electricity shown in the following Fig. 2, with the change of electricity trend in different time periods

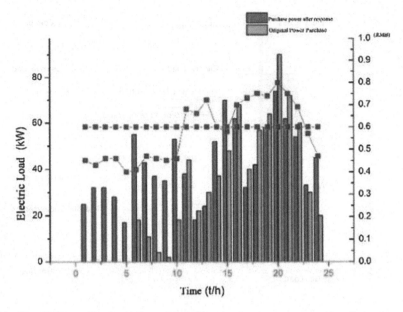

Fig. 2 Electricity price and load change analysis

throughout the day, the real-time electricity price is adjusted, and the purchased power after response is generally smaller than the original purchased power.

Shown in the following Fig. 3, Combined with the adjustment of the real-time electricity price, in the same time period: when the time-sharing price is lower than the reference price and the real-time price is lower than the reference price, the purchasing volume increases gradually after the response, and is lower than the original purchasing volume; The hourly price is higher than the reference price, and the real-time price is higher than the reference price. After the response, the purchasing volume decreases gradually and is lower than the original purchasing volume. The time-sharing price is lower than the reference price, and the real-time price is higher than the reference price. After the response, the purchasing volume increases gradually and is higher than the original purchasing volume. The hourly price is higher than the reference price, and the real-time price is lower than the reference price. After the response, the purchasing volume decreases gradually and is lower than the original purchasing volume. Thus, it can be concluded that the real-time electricity price is highly correlated with the load, and can be used as an important factor for Load Prediction, while the time-sharing price is less correlated with the load, and can be used as a secondary factor for Load Prediction.

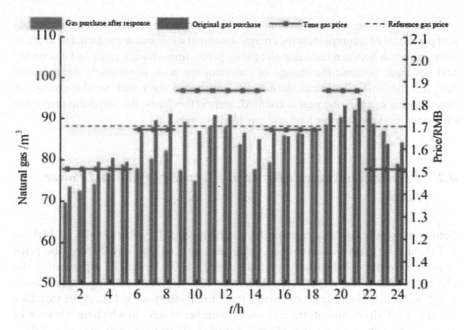

Fig. 3 Gas price and gas volume change analysis

4 Correlation Analysis Between Electric Price and Electric Gas Load

From the above price guidance analysis, we can see that besides meteorology and time, the economic adjustment factors of comprehensive energy price also affect the load. In order to illustrate the correlation between comprehensive energy price and load, this paper uses mutual information to make judgment. Mutual information is a measure of the degree of interdependence between two random variables. It can measure not only the linear relationship between the data characteristics of variables, but also the non-linear relationship [11]. It is derived from information entropy. The greater the mutual information between two variables, the stronger the correlation between them. The mutual information I (X, Y) formula for the two variables X and Y is as follows.

$$I(X, Y) = \sum_{i=1}^{m} \sum_{j=1}^{n} p(x_i, y_j) log_2 \frac{p(x_i, y_j)}{p(x_i)p(y_j)} \tag{1}$$

In the formula, m and N represent the number of elements in the variable X and Y, p(xi) represents the probability that element Xi in the variable X appears in all elements, p(yi) represents the probability that element Yi in the variable Y appears in all elements, and p(xi, yi) represents the probability of the joint distribution of variable X and Y.

As shown in the previous section, electricity purchase data, electricity price data and price data of a comprehensive energy demonstration area are selected to analyze the correlation between real-time electricity price, time-sharing price and electricity and gas load. Because the change of time-sharing price is relatively fixed in one day, in order to facilitate calculation, the number of days with similar change of time-sharing price in one year is counted, and on this basis, the correlation between electricity price and power load and gas load is analyzed.

4.1 Correlation Analysis Between Electricity Price and Power Load

Setting the power load sequence for day I indicates that $Li = \{li(1), li(2),\ldots, li(t),\ldots, li(T)\}$, and $li(t)$ represents the load value at time t on that day. Setting the price sequence for day I indicates $Ai = \{ai(1), ai(2), ai(3),\ldots, ai(T)\}$, representing the electrical value of the day's T sampling intervals. The data sampling interval is set to 0.5 h, $T = 48$. Formula (1) calculates the mutual information I (L, A) of variables L I and A I, P (li) represents the ratio of the number of days in which the value of Li load occurs to the total number of days sampled, P (ai) represents the ratio of days in which the value of AI electricity occurs to the total number of days sampled, and P (li, ai) represents the ratio of days in which both Li and AI occur in the sampling load to the total number of days sampled. To make the results easier to see, normalize the results to [0, 1]. The results showed that the days with I(L, A) > 0.6 were 253, accounting for 69.3% of the total days sampled. The number of days with I(L, A) > 0.5 was 274, accounting for 75.1% of the total days sampled. Figure 4 below shows the power price and load curve for a random day of data sampling. It can be seen that the trend of electricity price and power load change is similar. Therefore, the results of mutual information and price load curve show that the change of electricity price affects the trend of power load to a certain extent.

4.2 Correlation Analysis Between Electricity Price and Gas Load

As with the above steps, $Lj = \{lj(1), lj(2), \ldots, lj(t), \ldots, lj(T)\}$, $lj(t)$ represents the gas purchased at T-hour on that day. According to the standard for converting natural gas to electric energy, a cubic meter of natural gas equals about 10 degrees of electricity. Due to the inconsistency between natural gas price and electricity price, a j-day price difference sequence is set to represent the difference between real-time electricity price and time-of-use gas price for T sampling intervals on the same day, with $B_j = \{b_j(1), b_j(2), b_j(3),\ldots, b_j(T)\}$.

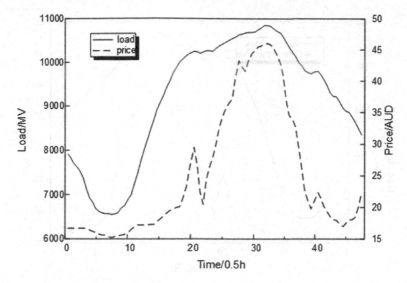

Fig. 4 One day's electricity price and load curve

$$b = b_p - \frac{b_e}{10} \tag{2}$$

where b_p for electricity price, b_e for gas price

The data sampling interval is set to 0.5 h, T = 48. Formula (1) calculates the mutual information J (L, A) of variables L_j and A_j, P (l_j) represents the ratio of the days of L_J purchasing value to the total days of sampling, P (a_j) represents the ratio of the days of A_j difference value to the total days of sampling, and P (l_j, a_j) represents the ratio of the days of both L_j and A_j occurring in sampling purchasing gas to the total days of sampling. Normalization is also used. The results showed that the number of days with I(L, A) > 0.6 was 226, accounting for 61.9% of the total days sampled. The number of days with I(L, A) > 0.5 was 254, accounting for 69.6% of the total days sampled. Figure 5 below shows the difference value and purchasing volume curve for a random day in a data sample. It can be seen from this that the variation trend of difference value and gas purchases is similar, and because the time-sharing price does not change much, the change of electricity price affects the trend of enough gas purchases to a certain extent.

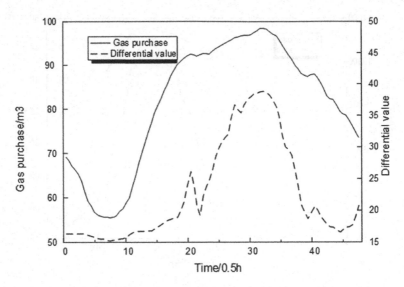

Fig. 5 One day electricity price and gas curve

5 Integrated Energy Load-Prediction Model Based on ATTENTION-LSTM

5.1 LSTM *NNs*

LSTM, as a special **RNN,** can avoid the problem of gradient disappearance and gradient explosion [12]. **LSTM** has a strong memory, can learn the long-term and short-term dependence information of time series, and can find the regular information from the historical data when predicting load. **LSTM** does this by adding "gate" information to neurons, and the structure is shown in the following Fig. 6.

LSTM's network structure is much more complex than **RNN**'s. Microscopically, **LSTM** add three gates, i.e., input i_t, forgotten f_t, and output o_t, with values ranging from [0, 1]. The input gate is primarily used to judge which properties are updated and what the new properties are about. The forgotten gate is the state information that was not used before. The output gate determines output content. These gates are determined by the output of the last unit h_{t-1} and the input of the current moment x_t. The candidate state values g_t of i_t, f_t, o_t and current neuron A are as follows

$$f_t = \sigma\left(W_{fx}x_t + W_{fh}h_{t-1} + b_f\right) \tag{3}$$

$$i_t = \sigma(W_{ix}x_t + W_{ih}h_{t-1} + b_i) \tag{4}$$

$$o_t = \sigma(W_{ox}x_t + W_{oh}h_{t-1} + b_o) \tag{5}$$

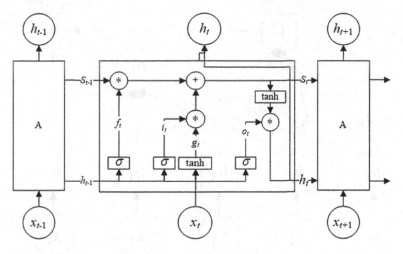

Fig. 6 Neuron structure of LSTM

$$g_t = tanh(W_{gx}x_t + W_{gh}h_{t-1} + b_g) \qquad (6)$$

W_{fh}, W_{ih}, W_{oh}, W_{gh}, W_{fx}, W_{ix}, W_{ox}, W_{gx} denotes the matrix weight obtained by multiplying the corresponding gate by the output h_{t-1} of the previous cell and the input X_t of the current moment. b_f, b_i, b_o, b_g are offsets, σ is sigmoid functions, and activation functions are tanh. The S_t is determined by the state value S_{t-1} of the last moment, the forgotten gate f_t, the input gate i_t and the candidate state value g_t of the current moment. When the new state value S_t is obtained, the output value h_t is expressed as the following formula:

$$S_t = f_t \cdot S_{t-1} + i_t \cdot g_t \qquad (7)$$

$$h_t = o_t \cdot tanh(S_t) \qquad (8)$$

In Formula, · represents the bitwise product of the elements in each multiplication vector.

5.2 Attention Theory Mechanism

Attention mechanism is a brain signal processing mechanism that simulates what human vision holds. Specific steps: Scan the information quickly, find the focus or target area it needs, ignore the useless information, and then focus more attention on the focus [13]. The basic idea of Attention mechanism is to be able to filter useful information, find important influencing factors by giving different weights to input

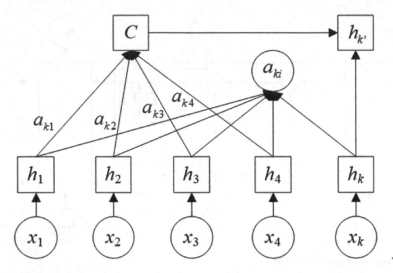

Fig. 7 Unit structure of Attention

sequence characteristics, and ignore unrelated factors, so as to improve the efficiency and accuracy of information processing. Importing Attention mechanism into LSTM to improve LSTM Load Prediction. the structure is shown in the following Fig. 7.

In the figure above, x_1, x_2,..., X_k is the input sequence, h_1, h_2,..., H_k is the state of hidden layer corresponding to input sequence x_k; a_{ki} is the weight of attention assigned to the current input by the hidden layer state feature h_k; $h_{k'}$ is the eigenvector value of the final output. The formula for calculating the **Attention** mechanism is as follows:

$$a_{ki} = \frac{exp(S_{ki})}{\sum_{j=i}^{T_x} exp(S_{kj})} \tag{9}$$

$$S_{ki} = vtanh(Wh_k + Uh_i + b) \tag{10}$$

$$C = \sum_{i=1}^{T_x} a_{ki} h_i \tag{11}$$

$h_{k'}$ is expressed as

$$h_{k'} = H(C, h_k, x_k) \tag{12}$$

5.3   Load Data Preprocessing

In order to facilitate the later data processing, the power load and the gas load are classified as a unified load. During the process of load data collection, due to various factors, data loss and data record errors may occur. If these data are used for model Load Prediction, large errors will inevitably occur. In order to improve the accuracy of **LSTM** model for Load Prediction, the original data needs to be preprocessed. Identify and correct anomalous data, and supplement missing data. Load data is sequential, usually continuous and smooth. For the load value at a certain time, the load value before and after it does not change much. If the change is too large, it is considered an outlier. In this case, the horizontal processing method of data processing is used, and the calculation formula is as follows.

If

$$max[|Y(d,t) - Y(d,t-1)|, |Y(d,t) - Y(d,t+1)|] > \varepsilon(t) \qquad (13)$$

Then

$$Y(d,t) = \frac{Y(d,t-1)+Y(d,t+1)}{2} \qquad (14)$$

In Formula, Y(d, t), Y(d, t + 1), Y(d, t-1) represent the power load values at t, t + 1 and t-1 on the day d, respectively. ε (t) is the threshold value.

Load data is similar to the same time of historical load, so there should not be too big difference between the value of load at one time on weekdays and rest days and the value at the same time on adjacent or rest days. If there is a big difference between load values, it is considered as an outlier. At this time, the vertical processing method is used to correct the difference, and the calculation formula is:

If

$$|Y(d,t) - m(t)| > r(t) \qquad (15)$$

Then

$$Y(d,t) = \begin{cases} m(t) + r(t) & Y(d,t) > m(t) \\ m(t) + r(t) & Y(d,t) < m(t) \end{cases} \# \qquad (16)$$

In the formula, m(t) is the average value of the load at t-moment in recent days, and r(t) is the threshold value. In addition, for the missing values of the original load data, this paper uses the clustering method to select the three most similar types of days, and then the calculated load values at each time are filled with the corresponding average values at the corresponding three-day corresponding time. Finally, the data is normalized using the **min–max** normalization method, ranging from [0, 1] to the following formula:

$$x^* = \frac{x - x_{min}}{xmin_{max}}\#$$ (17)

In the formula, x^* is the normalized value, x is the original data value, xmax is the largest sample value in the data, and xmin is the smallest sample value in the data.

5.4   Attention-LSTM *Short-Term Load Prediction Model*

Attention-LSTM is mainly composed of four parts, which are Input Vector, **LSTM** Hidden Layer, Attention Mechanism Layer, Full Connection Layer and Output Prediction Value. Input vectors enter the **LSTM** hidden layer from the input layer and are trained by at least four **LSTM** hidden layers. The trained vectors are output to the **Attention** layer, through which the weight vectors of the features are calculated. Then, the weight vectors are merged with the input vectors into the **Attention** layer, and a new vector is obtained as the input to the fully connected layer. After training in the fully connected layer, the final result is predicted and output. The **Attention-LSTM** prediction model is shown in the following Fig. 8.

In the input vector, price factor, the following 11 feature factors are selected as input features, expressed as $U_i(t) = [p_i(t), v_i(t), o_i(t), y_i(t), c_i(t), h_i(t), r_i(t), L_{i-1} (t-1), L_{i-1} (t-1), w_i(t), f_i(t)]$ as shown in the following Table 1. Holiday factors are discretized in the treatment, $w_i(t) = 0$ on weekdays and $f_i(t) = 1$ on holidays.

The main function of the **LSTM** layer is to extract valuable information through continuous learning and training, and to forget and discard useless information. Theoretically, the deeper the number of layers, the stronger the ability to fit the function in theory, and the better the result is theoretically. However, in fact, the deeper the number of layers may cause problems of over-fitting, but it also increases

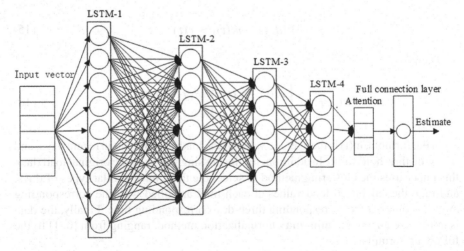

Fig. 8 Attention_LSTM model

Table 1 Selected feature factor

influence factor	Feature representation	Feature description
Economic factors	$p_i(t)$	Peak hour electricity price
	$v_i(t)$	Valley-hour electricity price
	$o_i(t)$	Peacetime electricity price
	$Y(t)$	Gas price
Meteorological factors	$c_i(t)$	Predicted temperature at prediction point time
	$h_i(t)$	Predicted humidity at prediction point time
	$r_i(t)$	Rainfall likelihood at prediction point time
Historical data	$l_{i-1}(t)$	Load value at the same time the day before forecast point
	$l_{i-1}(t-1)$	Load value one day before forecast point
Holiday factor	$w_i(t)$	Working day
	$f_i(t)$	Holiday and vacations

the difficulty of training and makes the model difficult to converge. Therefore, a suitable time and good training results should be selected. The model establishes four **LSTM** layers experimentally. Generally, the number of nerve cells in each layer of the model is 2^n, so the number of neurons in the first layer of training should not be too small, otherwise the training effect cannot be guaranteed. The first layer is 512, the second layer is 256, the third layer is 128, and the fourth layer is 64. The main reason that the number of neurons decreases in turn is that a large number of neurons can learn many lower-order features in the early stage, and the lower-level feature feeds and a small number of neurons can train to obtain higher-order features in the late stage to reduce data interference. In addition, the benefit of full-layer compression is that as data increases, the parameters of full-layer compression multiply, so proper compression is important before you enter full-layer compression.

Attention layer mainly applies the learned feature weights to the input vectors of the next time step, reflecting the effect of key features on the prediction results. The data enters the full connection layer from the Attention layer to get the load value of the prediction point. This model sets the training step size to 11.

6 Example Analysis

6.1 Experimental Data

Data from 2017 to 2019 were taken as training set, and data from 2020 to 2021 were taken as test set. The sample interval of electric power load and gas data collection is 5 min. Data content: sampling time point, gas consumption and load value. Meteorological data collection is sampled at one minute intervals, data content: sampling time point, weather condition, temperature and humidity, wind level, light intensity. The selected data characteristic factors are shown in Table 1.

6.2 Experimental Environment

The experimental hardware environment uses a graphical workstation. Processor is Intel XEON, 8 cores 16 threads, 3.6 GHz, memory capacity is 64G, solid state hard disk 1 T, graphics card is 8G unique. The software architecture uses the in-depth learning framework Tensorflow.

6.3 Contrast Index

Three indicators are selected as the evaluation criteria for this experiment, which are mean absolute percentage error (MAPE), root mean square error (RMSE), and prediction accuracy (FA). The smaller the MAPE and RMSE values, the better the model training effect. The larger the FA values, the better the model training effect. The calculation formulas for MAPE, RMSE and FA are as follows:

$$y_{MAPE} = \frac{1}{n} \sum_{i=1}^{n} \left| \frac{l_{real}(i) - l_{fore}(i)}{l_{real}(i)} \right| \tag{18}$$

$$y_{RMSE} = \sqrt{\frac{\sum_{i=1}^{n} \left(l_{real}(i) - l_{fore}(i) \right)^2}{n}} \tag{19}$$

$$y_{FA} = \left[1 - \frac{|l_{real}(i) - l_{fore}(i)|}{l_{real}(i)} \right] \times 100\% \tag{20}$$

n is the total number, $l_{real}(i)$ represents the true value at time i, and $l_{fore}(i)$ represents the predicted value at time I.

6.4 Experimental Results and Analysis

The experiment is divided into three parts. The first part is to determine the number of hidden layers for Attention-LSTM model training, mainly through y_{MAPE} size judgment; The second part verifies that electricity price and gas price have certain influence on Load Prediction. Adding electricity price and gas price to the characteristic factor can improve the accuracy of Load Prediction, mainly y_generated according to different input characteristics. y_{MAPE} and y_{RMSE} result value to verify;

Table 2 Different LSTM training layer load prediction results

LSTM layers	Training rounds	Training batch size	Eigenvalue	y_{MAPE}
2	600	1024	11	0.589
3	600	1024	11	0.577
4	600	1024	11	0.556
5	600	1024	11	0.554
6	600	1024	11	0.553

The third part verifies the validity of the proposed model by comparing it with other methods.

Determining Number of Training Layers for Attention-LSTM Model. In the form of control variables, the load prediction effect was tested by increasing the number of training layers of LSTM under the condition that the number of training rounds, the size of training batches and the number of features of the model were constant. The experimental results are shown in the following Table 2. The larger the number of LSTM training layers, the smaller the value of y_{MAPE}, and the model effect gradually becomes better. However, when the number of LSTM training layers is more than 4, y_{MAPE} changes slightly with the increase of the number of layers. Therefore, 4 training layers are selected.

Validation of influence of electricity price characteristics on Load Prediction. In order to explain that electricity price factors have a certain influence on power load prediction, the characteristic factor number of input vector of **Attention-LSTM** model is adjusted to some extent. **CH** refers to the feature vectors of normal price, valley price and peak price factors removed from Table 1; **CHP** refers to the feature vectors of peak price added on the basis of **CH**; **CHPV** refers to the feature vectors of valley price added on the basis of **CHP**; **CHPVO** refers to the feature vectors containing all factors in Table 1. Table 3 shows the load prediction results of attention-LSTM model through different input feature vectors. It can be seen that the values of y_{MAPE} and y_{MAPE} when **CH** is used as feature vector are all lower than those of **CHP**, **CHPV** and **CHPVO**, and the prediction accuracy is the lowest, indicating that consideration of electricity price characteristics can improve the accuracy of model load prediction and positively affect load. In addition, the prediction effect of **CHP** is lower than that of **CHPV**, which indicates that the more detailed decomposition of electricity price features can effectively improve the accuracy of model prediction. Meanwhile, it also indicates that when all the features in Table 1 are used as input, the model can fully learn useful information and rules in the data, and the prediction effect is the best.

Validation of Attention-LSTM Model. The short-term Load Prediction methods based on **RF**, **RNN** and **LSTM** are selected for comparison experiment. Short-term Load Prediction for the first half year and one random day in the first half of 2017 is carried out continuously, and the results are shown in Table 4. The results show that **Attention-LSTM** model has the smallest y_{MAPE}, $y_{FA,avg}$ in several models, and the prediction results are better than other models. Among them, $y_{FA,avg}$ are the average of

Table 3 Prediction results for different input vectors

Model input Eigenvector	2020 year		2021 year	
	y_{MAPE}	y_{RMSE}	y_{MAPE}	y_{RMSE}
CH	0.907	8.725	0.994	9.012
CHP	0.892	7.842	0.701	8.711
CHPV	0.645	7.489	0.598	8.546
CHPVO	0.573	6.637	0.586	7.939

Table 4 The result of model prediction

Model	one day		Half a year	
	y_{MAPE}	$y_{FA,avg}$	y_{MAPE}	$y_{FA,avg}$
Random forest	0.944	96.81	1.301	95.32
RNN	0.885	96.85	0.994	97.12
LSTM	0.801	97.73	0.989	98.19
Attention-LSTM	0.645	98.86	0.678	98.48

prediction accuracy. From the above results, the **Attention-LSTM** model is superior to other models in both validity and accuracy.

Figure 9 shows y_{MAPE} value of absolute percentage error of load prediction of four models in 24 h on a certain day, which is calculated every one hour. It can be seen from this that the y_{MAPE} value of **Attention-LSTM** model is the best overall result, much lower than other methods except that it is slightly larger at individual points. Although **LSTM** and Random forest models achieved low y_{MAPE} value at the beginning, the subsequent effect gradually deteriorated with the increase of time, while **RNN** model only performed well at some points, and the overall prediction accuracy fluctuated greatly. Figure 10 shows the comparison results of the predicted value and the real value of the load prediction of the four models on a certain day, with sampling every 15 min. As can be seen from the results, **Attention-LSTM**, **LSTM** and random forest model all perform relatively well when sampling points are small. However, with the increase of sampling points, only **Attention-LSTM** model has the closest prediction result to the real value, and its prediction accuracy is better than other models. It is proved that the **Attention-LSTM** model is effective and accurate in short-term Load Prediction.

7 Conclusion

In this paper, the mutual information algorithm is used to verify that the economic adjustment factors of electricity price have some influence on the power load and

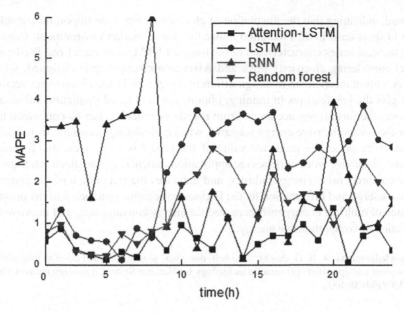

Fig. 9 Absolute percentage error comparison of Load Prediction for one day

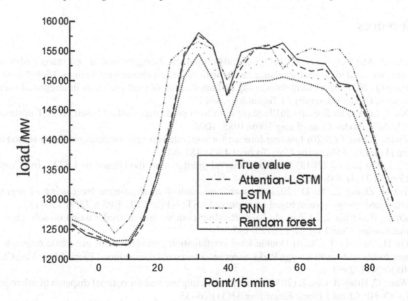

Fig. 10 Comparison of the predicted load value of a certain day with the true value

gas load, indicating that the fluctuation of electricity price is an important consideration in short-term Load Prediction under the power market environment. Considering the time series characteristics of the load, a **LSTM** short-term Load Prediction model considering electricity price and **Attention** mechanism is designed, which uses Attention mechanism to assign different weights to hidden layer eigenvectors, highlights the key features in training, obtains the final Load Prediction value, and improves the forecasting accuracy. From the above model, it can be concluded that under the comprehensive energy scenario, when considering economic factors such as electricity price, the predicted value of the model is more accurate. Technical aspects: The above work enriches the application dimension of artificial intelligence in the comprehensive energy industry, and enhances the integration of the comprehensive energy and information fields; Economic benefits: conducive to the optimal dispatch of multiple energy resources, reduce energy consumption, and improve the utilization of comprehensive energy.

Acknowledgements R. B. G. thanks "One belt, one road" jointly build and develop joint energy management and operation optimization technology for National Smart City energy network (Item No:2020YFE0200400).

References

1. Liao Z, Wan W, Chen X (2021) Day-ahead optimal scheduling method of park integrated energy system considering price guidance mechanism. Southern Power Syst Technol 15(9):53–68
2. Fang Q (2021) Short term forecasting and application of multiple loads in integrated energy system. Dalian University of Technology, vol 1
3. Xu Y, Lu Y, Zhu B et al (2019) Short term load prediction method based on FFT optimized RESNET model. Control Eng 26(6):1085–1090
4. Zhang T, Sun T (2020) Loading forecast for integrated energy system considering season and trend factors. J Shenyang Univ Technol 42(5):481–487
5. Yu J, Bao Z, Li Z (2018) User load interval prediction method based on LSTM. Ind Control Comput 31(4):100–102
6. Tia H, Zhang Z, Yu D (2021) Research on multi-load short-term forecasting of regional integrated energy system based on improved LSTM. Proc CSU-EPSA 33(9):130–137
7. Zhang H, Zhang P, Li Z et al (2018) Reading comprehension model based on self-attention mechanism. Chin J Inf 32(12):129–135
8. Gu H, Yu J, Li Y (2020) Double-level combination optimization of economic dispatch for new towns with multi-energy Parks under environmental constraints. J Electr Electr Eng China 40(8):2441–2453
9. Wang C, Hong B, Guo L (2013) General modeling method for optimal dispatch of microgrids for COHP. China J Electr Electr Eng 33(31):26–33
10. Sun Q, Xie D, Nie Q (2020) Study on optimal economic dispatch of comprehensive energy system in parks with electric-thermal-cold-air load. China Electric Power, 53(4):79–88
11. Wang Y, Li J, Zeng H et al (2019) A real-time feature extraction algorithm based on mutual information. Minicomput Syst 53(4):1242–1247
12. Abdel-Nasser M, Mahmoud K (2017) Accurate photovoltaic power forecasting models using deep LSTM-RNN. Neural Comput Appl 10:1–14
13. Wang Y, Zhong H, Yu N et al (2019) Resident user non-invasive load decomposition based on seq2seq and Attention mechanism. China J Electr Eng 39(1):75–83

Design of Concentrated Solar Power Plant with Molten Salt Thermal Energy Storage

Kashif Ali and Jifeng Song

Abstract The use of mirrors and Concentrated Solar Power (CSP) allows us to harness the energy for our own use. In 2032, the development of CSP is predicted to increase by 34%. Focusing the sun's heat onto a receiver, CSP systems convert it into heat. The steam is then used to power a turbine that generates energy. Concentrated solar power, when used in conjunction with other sources of energy, can help to improve the reliability of the electricity grid. The aim of this paper is to Design a CSP plant with molten salt thermal energy storage. A 70 MW CSP plant is designed with parabolic collector. MATLAB is software used for simulation of plant. The results of model shows that the overall generation of system 70 MW when adding molten salt storage, it increases efficiency of system and provide additional power 2 MW to grid. CSP generation increased with molten salt storage.

Keywords CSP · Molten salt · Energy storage · Efficiency

1 Introduction

With the aid of mirrors, solar concentrated power can be used to generate power. For example, the sun's rays can be focused and concentrated on a certain area by using mirrors. As a result, a turbine generates electricity by turning the steam into electricity. To keep the process running indefinitely, CSP technology uses heat storage technology. When the sun isn't out, it can still be used before and after sunrise or sunset. CSP generation is predicted to rise by 34% in 2019 predicted by the International Energy Agency (IEA). A lot of things have worked out for CSP in the past, but it has to go before it meets its development goals of sustainability. These goals call for an average annual growth rate of 25% after 2030 [1, 2]. Solar concentrated power systems direct the sun rays toward a receiver, where they are transformed into heat (see in Fig. 1). As a result, a turbine generates power from the steam. Power can be stored for periods of low sunlight at CSP installations using

K. Ali · J. Song (✉)
North China Electric Power University, Beijing, China
e-mail: Songjifang@ncepu.edu.cn

Fig. 1 Parabolic collector for concentrated solar power plant (CSP)

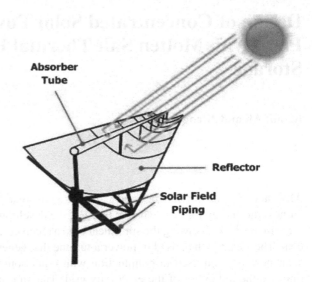

thermal energy storage devices. CSP is a useful renewable energy source because of its ability to store energy. In order to create hybrid power plants, CSP systems can be used in conjunction with other types of electricity generation. It is possible to combine CSP with coal-burning, biofuel, and natural gas power facilities [3]. CSP's most evident benefit is that it is a renewable energy source. With an endless supply and endless use, it is a long-term energy source. As a bonus, it minimizes greenhouse gas emissions. CSP is more environmentally friendly than fossil fuels since it uses the earth's natural resources instead of emitting CO_2. Improved air quality and delayed global warming are possible outcomes. A continuous supply of electricity is provided by CSP compared to wind power and solar photovoltaics (PV).

Because CSP systems use molten salts to store solar energy, its output is consistent and predictable. Converting existing steam-based power facilities to CSP is a simple process. Cars fueled by fossil fuels can use CSP systems. In comparison to hydrocarbon-based and nuclear reactors, CSP plants have lower operating costs since maintenance and operations are easier [4–6]. Concentrated solar power can aid in the development of a more reliable power system when used in conjunction with other sources of energy. CSP can help meet future electricity needs if it is added to the energy mix. It can also help with making it easier to pump out of the ground and oil recovery by concentrating heavy oil. A portable power supply is also possible with this device. According to Ecofys, a renewable energy firm, CSP might be utilized to produce cost-effective hydrogen that could be used to power transportation [7–10].

Arid, or "sun belt," regions are the most typical locations for CSP plants since they require a lot of space. When the thermal cycle is completed, the cooling towers must be wet and the mirrors must be cleaned and maintained with water. As the demand for water grows across Middle East and the, the North Africa ability of these countries to consume huge amounts of fresh water is sometimes questioned. Another effect is the environmental impact of CSP plants in terms of their visual appearance.

A lot of people don't like how CSP projects make it hard to see natural landscapes. Ecosystems and communities take longer to recover from problems in the sun belt because of its dry nature. This is because the sun belt is important for the environment and serves as a safe haven for endangered species, even though there is no farming in this area [11].

There are a number of variables that affect the effectiveness of CSP systems. The concentrated solar power system efficiency depends on the type of system, the receiver, and the engine. An energy Sage study found that efficiency of most CSP systems ranges from 7 to 25%. Wind turbines, on the other hand, can attain efficiencies of up to 59% with hydropower systems. The efficiency of most photovoltaic solar panels ranges from 14 to 23%, which is comparable to that of concentrated solar power plants [12, 13].

Using a grid-tied inverter to convert alternating current (AC) from direct current (DC) can lead to energy losses of 4–13%. These energy losses can be blamed on the conversion of AC from DC. CSP is preferred by power networks because the steam engine turns heat energy from electrical energy to mechanical energy, which is the main source of power generation, which is why it is preferred. People can connect CSP directly to the grid and have backup power from combustible fuels because the heat engine doesn't have a problem with converting energy between different types of fuel [14]. A lot of other things make CSP better than PV, like being more efficient and making more money, having a built-in thermal storage capacity that can be used to make power, and being better at hybrid operations. CSP can keep making electricity even when the sky is covered in clouds or it's dark outside. Simulating and modelling the parabolic trough collector in this paper is what we're going to talk about here. In order to show off new renewable energy research and teach students how to use renewable energy technology, a solar park was built on Spectrum Blvd at USF Tampa [15–17]. A lot of electricity is made at this plant, and both USF and the electric grid benefit from it. The Daily Integration (DI) method was used to figure out the average DNI for the solar plant's site (FL, Tampa, USF). In Tampa, the average DNI for the whole year is 4.6 kWh/m^2-day, which is the direct normal solar radiation for the city. It was based on these solar radiation measurements and the analysis of solar shading for rows of solar collectors to figure out how much sun each row got. Construction of a 50 kW CSP system with a solar field layout. Sopogy Inc.'s parabolic trough collectors were used in the sun field to get 430 W/m^2 of thermal energy after losing some of it. Phase change materials (PCMs) are used in the concentrated solar power storage system, which is meant to work with the solar field [18–20].

2 Methodology

The thermal radiation parameters of the solar receiver and the thermal radiation features of the heat engine govern the efficiency with which incident solar energy is converted into mechanical work (see Fig. 2). When the heat engine, which uses

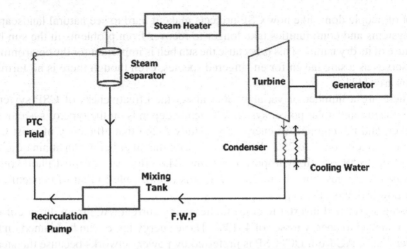

Fig. 2 Flow diagram a process of parabolic trough controller (PTC) based CSP

Carnot's principle, is paired with the solar receiver's efficiency, the result is a combination of heat and electricity. The eta receiver then converts the heat into mechanical energy. With the help of a generator, a machine converts electrical energy from mechanical energy. The overall efficiency of a solar receiver and a mechanical converter can be calculated as follows:

$$\eta = \eta_{reciever} \times \eta_{mechanical} \times \eta_{generator} \times \eta_{optics} \tag{1}$$

where η is the efficiency of the project and the efficiency of converting mechanical energy from heat energy called mechanical efficiency, Generator denotes the efficiency of converting electrical power from mechanical energy and he efficiency of converting heat energy into optics is called optics efficiency.

$$\eta_{reciever} = \frac{Q_{absorbed} - Q_{lost}}{Q_{incident}} \tag{2}$$

The fluxes received and lost by the solar system and the incoming solar flux and receiver are denoted as absorbed, lost, and incident, respectively. To make up for these losses, conventional engines have real world efficiencies that range from 51 to 71% of Carnot efficiency. This is because of things like windage and heat loss in the moving parts.

2.1 Modelling

Figure 2 depicts a Rankine cycle in the form of a circle with inputs and reference values. To begin, the heat transport in each component of the cycle is analyzed using a theoretical concept. These points of reference can be used to calculate the condenser water flow rate, steam flow rate and other related parameters.

Here are the performance evaluation formulae for the entire system. The computations are based on assumptions. It is assumed that the system is in a stable state. Pressure changes are modest everywhere except for turbines and pumps. The formula for estimating the collector's useful energy rate is as follows:

$$Q_{gain} = m_{cl}(h_7 - h_6) = \eta \cdot I \cdot A_p \tag{3}$$

$$\eta = \eta_0 - U_L \cdot \left(\frac{\Delta T}{I}\right) \tag{4}$$

$$\Delta T = T_m - T_a \tag{5}$$

where if the Feed Water Pump (F.W.P) and work done by the Recirculation Water Pump (RC.W. P), as well as the heat lost through the pipes, you get AP, which is the area of the collector in square meters. UL is the loss co-efficient based on the area (W/m²). mcl is the mass flow rate of steam through the Collector (kg/s), I direct normal incidence radiation W/m² and; $T_m = (T^6 + T^7)/2$, T_a is ambient temperature (C);

$$Q_{gain} = m_{CL}(h_7 - h_6)(h_1 - h_3) \tag{6}$$

$$m_{CL}(h1 - ha) = \eta \cdot I \cdot A_p \tag{7}$$

The PTC's mass flow rate (m_{cl}) of the inlet will usually stay the same during operation. The mean temperature difference (ΔT) at part load and will change the dryness fraction (ΔT) of the PTC exit and we can figure out TT's value if we don't think about pump effort and assume that the PTC field design point outlet dryness fraction (X_7). There are two things you can get from PTC field pressure and enthalpy that can be used to figure out the temperature (T_7) of the PTC field.

$$h_6 = (1 - x_7) \cdot h_8 + x_7 \cdot h_3 \tag{8}$$

As previously stated, Ti and Hi denote the enthalpy and temperature at the i-th state point respectively, the ability to see is defined as the optical efficiency o,

$$\eta_0 = \rho_c \gamma \tau \alpha K_\gamma \tag{9}$$

The transmission, incidence angle modifier of the mirror's intercept factor, absorbance, and reflectance, glass cover are all shown:

$$A_p = (w - D_{CO})L \tag{10}$$

The radiation heat coefficient among the cover and the environment is determined as follows;

$$U_L = \left(\frac{A_r}{(h_{c,ca} + h_{r,ca})A_C} + \frac{1}{h_{r,cr}}\right)^{-1} \tag{11}$$

The Stefane-Boltzmann constant is; $\varepsilon_{C\nu}$ is the emittance of the cover. Radiation heat coefficient can be expressed in two ways: as a percentage or as a number.

$$h_{r,ca} = \varepsilon_{C\nu}\sigma(T_c + T_a)(T_{c^2} + T_{a^2}) \tag{12}$$

The Stefane-Boltzmann constant is $\varepsilon_{C\nu}$ is the cover's emittance. The heat radiation coefficient between the cover and the receiver.

$$h_{r,cr} = \frac{\sigma(T_c + T_{r,av})(T_c^2 + T_a^2)}{\frac{1}{\varepsilon_r} + \frac{A_r}{A_c}(\frac{1}{\varepsilon_{cv}} - 1)} \tag{13}$$

where ε_r is the emittance of the receiver and av the average value, The coefficient of cover-to-ambient convection heat loss is defined as;

$$h_{c,ca} = \left(\frac{Nuk_{air}}{D_{c.o}}\right) \tag{14}$$

'K_{air} measurement of air's thermal conductivity, and Nu its Nusselt number. The subscript r denotes the receiver. Here's a calculation for calculating how hot the cover will be when it's open:

$$T_c = \frac{h_{r,cr}T_{r,a} + (h_{c,ca} + h_{r,ca})T_O\frac{A_r}{A_c}}{h_{r,cr} + (h_{c,ca} + h_{r,ca})\frac{A_r}{A_c}} \tag{15}$$

The amount of solar radiation that reaches the collector and is converted to heat before entering the system is measured.

$$Q_{solar} = m_{cl}.(h_7 - h_6m + (h_1 - h_9) \tag{16}$$

The performance equations for the entire system will be examined next. The output of a steam turbine is defined as:

$$W_{st} = m(h_1 - h_2) \tag{17}$$

The enthalpy is h and steam are st, hence h = st. The net power of the steam Rankine cycle is given as:

$$W_{cyc} = \eta_g W_{st} - (W_{F,W,P} + W_{Rc.P.W})$$ (18)

$W_{F.W.P}$ is the power for the feed water pump, whereas $W_{RC.W.P}$ is the power for the recirculation pump.

$$W_{F,W,P} = m(h_4 - h_3) = m\vartheta(P_4 - P_3)$$ (19)

$$W_{Rc.P.W} = m(h_6 - h_5) = m\vartheta(P_6 - P_5)$$ (20)

The net electrical efficiency of the steam Rankine cycle system is calculated using the formula:

$$\eta_{el} = \frac{W_{cyc}}{Q_{solar}}$$ (21)

3   Simulation and Results

MATLAB Software used to evaluate the CSP Plant system's performance, many scenarios are simulated in this paper. It described that simulate 2 variations in solar irradiation and PV module operating temperature. (See Fig. 3) shows the irradiance of 1000 W/m^2 and Fig. 4. Shows 800 W/m^2 from solar panel with the duration of 5 s.

The irradiance shift from 1000 to 800 W/m^2 (See Fig. 4). As can be seen from the figure, the tracking accuracy degrades with decreasing irradiation. At lower irradiation levels, reduced duty ratio command step sizes may improve CSP plant accuracy. To accomplish maximum power, this irradiance level should be high for power plant meant that it will take longer.

The power total power flow of CSP plant at 1000 W/m^2 can be seen on Fig. 5. Also, figure shows that the power receives is 70 MW with the time of 5 s. The initially power was high then it stabilizes at 70 MW.

Figure 6 shows that the output power of plant. 70 MW of fix generation is produced by CSP plant. When molten salt storage is added to generation It shows that the storage is increased by 2 MW.

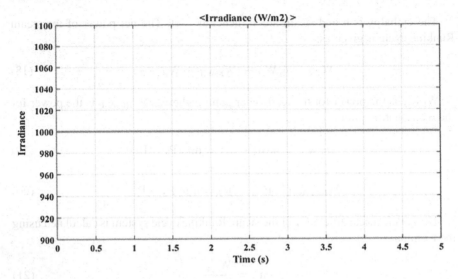

Fig. 3 Irradiance at 1000

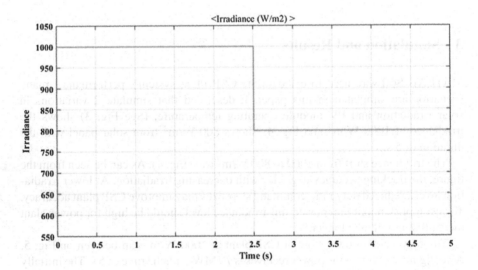

Fig. 4 Different irradiance level

4 Conclusion

This paper came to the conclusion that a solar concentrated power plant is a viable option for conceptual design calculations in this research. The MATLAB software is used to design model in order to predict the power, energy, irradiance and storage of the system. The overall generation of system 70 MW when adding molten salt

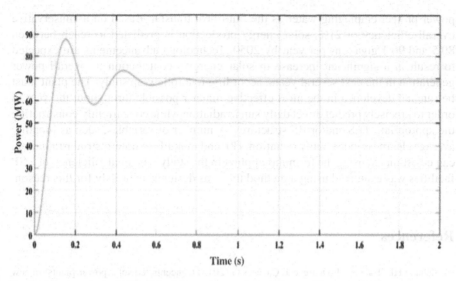

Fig. 5 Output power of CSP plant

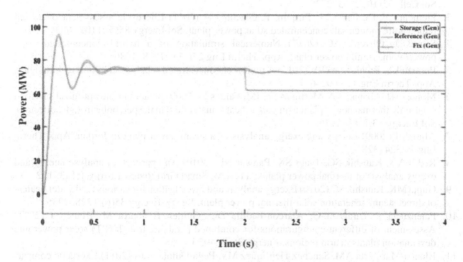

Fig. 6 Output power of power plant with Molten salt storage

storage, it increases efficiency of system and provide additional power 2 MW to grid. The influence of the solar field in ambient settings, as well as the previously mentioned solar radiation, are incorporated into the control theory for the turbine-generator unit. According to the graph, it is expected that heat gain will continue to increase indefinitely until the maximum power is given. These disclosures may aid in the establishment of a more effective control strategy and the better planning of power plant operations in the future. The modeling data for a 70 MW solar concentrated power project employing water as the fluid heat transfer. Based on a conservative overall efficiency of 21%, solar energy production is predicted to reach between 80.5 and 90.1 giga tons per year by 2050. Technology advancements are expected to result in a significant increase in solar energy's contribution to overall power generation in the next several years, according to a follow-up study. The plant must be run and developed in the most effective manner possible throughout the year in order to precisely predict direct daily sun irradiation while concurrently constructing the appropriate TES and/or BS structures. A number of variables, such as monthly average clearness index, daily radiation, H0, and monthly extraterrestrial irradiation, can be estimated using the formulas applied in the study. The total efficiency of CSP facilities was examined using a method that was designed expressly for this reason.

References

1. Zhang HL, Baeyens J, Degrève J, Cacères G (2013) Concentrated solar power plants: review and design methodology. Renew Sustain Energy Rev 1(22):466–481
2. Barlev D, Vidu R, Stroeve P (2011) Innovation in concentrated solar power. Sol Energy Mater Sol Cells 95(10):2703–2725
3. Piemonte V, De Falco M, Tarquini P, Giaconia A (2011) Life cycle assessment of a high temperature molten salt concentrated solar power plant. Sol Energy 85(5):1101–1108
4. Soares J, Oliveira AC (2017) Numerical simulation of a hybrid concentrated solar power/biomass mini power plant. Appl Therm Eng 25(111):1378–1386
5. Pramanik S, Ravikrishna RV (2017) A review of concentrated solar power hybrid technologies. Appl Therm Eng 127:602–637
6. Montes MJ, Abánades A, Martínez-Val JM, Valdés M (2009) Solar multiple optimizations for a solar-only thermal power plant using oil as heat transfer fluid in the parabolic trough collectors. Sol Energy 83:2165–2176
7. Aljundi I (2009) Energy and exergy analysis of a steam power plant in Jordan. Appl Therm Eng 29:324–328
8. Reddy VS, Kaushik SC, Tyagi SK, Panwar NL (2010) An approach to analyse energy and exergy analysis of thermal power plants: a review. Smart Grid Renew Energy 1:143–152
9. Gupta MK, Kaushik SC (2010) Exergy analysis and investigation for various feed water heaters of direct steam generation solar thermal power plant. Renew Energy 35(6):1228–1235
10. Palenzuela P, Zaragoza G, Alarcon-Padilla DC, Guillen E, Ibarra M, Blanco J (2011) Assessment of different configurations for combined parabolic-trough (PT) solar power and desalination plants in arid regions. Energy 36(8):4950–4958
11. Blanco-Marigorta AM, Sanchez-Henríquez MV, Peña-Quintana JA (2011) Exergetic comparison of two different cooling technologies for the power cycle of a thermal power plant. Energy 36(4):1966–1972
12. Xu C, Wang Z, Li X, Sun F (2011) Energy and exergy analysis of solar power tower plants. Appl Therm Eng 31:3904–3913
13. Kaushik SC, Reddy VS, Tyagi SK (2011) Energy and exergy analysis of thermal power plants: a review. Renew Sustain Energy Rev 15:1857–1872
14. Kumar KR, Reddy KS (2012) 4-E (energy–exergy–environmental–economic) analyses of line-focusing stand-alone concentrating solar power plants. Int J Low-Carbon Technol 7(2):82–96. https://doi.org/10.1093/ijlct/cts005
15. Reddy KS, Devaraj VA (2012) (4-E) Energy–exergy–economic–environment analysis of stand-alone solar thermal power plants and solar-coal hybrid plants. J Fundam Renew Energy Appl 2:1–6. https://doi.org/10.4303/jfrea/R120308

16. Reddy VS, Kaushik SC, Tyagi SK (2012) Exergetic analysis and performance evaluation of parabolic trough concentrating solar thermal power plant (PTCSTPP). Energy 39:258–273
17. Han W, Hongguang J, Rumou L, Qibin L (2014) Performance enhancement of a solar trough power plant by integrating tower collectors. Energy Procedia 49:1391–1399
18. Avezova NR, Khaitmukhamedov AE, Usmanov AYu, Boliyev BB (2017) Solar thermal power plants in the World: the experience of development and operation. Appl Sol Energy 53(1):72–77
19. Kuchkarov AA, Kholova ShR, Abdumuminov AA, Abdurakhmanov A (2018) Optical energy characteristics of the optimal module of a solar composite parabolic-cylindrical plant. Appl Sol Energy 54(4):293–296
20. Kariman H, Hoseinzadeh S, Heyns PS (2019) Energetic and exergetic analysis of evaporation desalination system integrated with mechanical vapor recompression circulation. Case Stud Therm Eng 16. Id 100548

Design of Biomass Straw Liquefaction Unit Based on Plasma Technology

Yike Zhang, Jinmao Li, Jiaqi Wang, and Shilong Cheng

Abstract To solve the problems of low comprehensive utilization rate of corn straw and environmental pollution caused by large-scale combustion, a technical scheme of treating biomass straw by plasma was proposed, and the plasma biomass straw liquefaction unit was designed and studied. The device includes: plasma power supply, gas mixer, multi-function feeder, plasma conversion reactor, detection device, cyclone separator, condenser, gas collection device, etc. After the corn straw is initially crushed by the grinder, it enters the plasma conversion reactor through the multi-functional feeder, and generates solid, liquid and gas substances that can be reused under the action of plasma flame. The device is energy-saving and environmentally friendly, in line with the development requirements of carbon peak and carbon neutral times.

Keywords Plasma · Plasma Power Supply · Environmentally Friendly · Carbon Peak · Carbon Neutral

1 Introduction

In recent years, plasma technology has developed rapidly. The unique chemical activity and high reactivity of plasma make it possible for many chemical reactions that are difficult or impossible to be realized by traditional methods. People began to try to apply plasma technology to biomass thermal conversion and utilization [1].

At present, energy and environmental issues have become international hot issues, attracting people's extensive attention [2]. With the huge global consumption of fossil raw materials and environmental pollution, the high-value utilization of biomass

Y. Zhang
Department of Electrical and Information Engineering, Heilongjiang University of Technology, Jixi, China

J. Li (✉) · J. Wang · S. Cheng
Heilongjiang Provincial Key Laboratory of Plasma Biomass Materials Research and Testing, Heilongjiang University of Technology, Jixi, China
e-mail: lijinmao000@163.com

renewable resources has become a research hotspot in the world today [3]. Internationally, studies on biomass gasification and liquefaction to produce liquid fuels have been carried out in depth, but the development of this technology lags behind in China [4, 5]. China rural areas produce more than 900 million tons of plant straw every year, with a huge yield. Biomass energy is abundant, but the direct utilization technology of straw is mainly incineration, and the research technology with high added value of straw develops slowly. The specific technology of comprehensive straw control is not perfect, policies and capital investment are insufficient, market operation is not strong enough, straw treatment equipment and related processing facilities are not perfect, straw utilization technology is not mature, and the efficiency and benefit of comprehensive straw utilization need to be improved. The rational transformation of biomass is imminent.

2 Comparison of Existing Biomass Straw Treatment Technologies

2.1 Mechanical Crushing

Mechanical crushing is to turn lignocellulose into particles by mechanical force. There are many mechanical crushing equipment and ways, such as knife grinding, disc grinding, steam grinding and ball grinding. Mechanical crushing has the advantage of simple and effective operation, but high energy consumption and high cost [6].

2.2 Biological Treatment

Lignin was decomposed by microorganism to remove its wrapping effect on cellulose. Commonly used microorganisms are white rot bacteria, brown rot bacteria and soft rot bacteria. In the process of culture, specific enzymes for the decomposition of lignin are produced, which reduces the generation of by-products. The most effective of these are white rot fungi, which secrete effective lignin-degrading enzymes such as peroxidase and laccase. Nowadays, the pretreatment of fiber raw materials by white rot bacteria has been widely studied. Although the biological pretreatment technology based on white rot bacteria can degrade and destroy the protective barrier of biomass raw materials and improve the conversion efficiency, the efficiency is relatively low and still needs to be improved, and it takes a relatively long time to use biological treatment [7].

3 Propose Technical Proposal

With the rise of temperature, the existence state of material generally presents the transformation process of solid state, liquid state and gas state. Solid, liquid and gas are the three states of material. When the temperature of gaseous substances rises to several thousand degrees, due to the intensified thermal movement of the molecules, the collision between them will ionize the gas molecules, and the substance will become a mixture of positive ions and electrons that move freely and interact with each other, namely the fourth state of the substance—plasma. The temperature ranges from 100 K for low temperature plasma to $10^8 \sim 10^9$ K for ultra-high temperature fusion plasma. The plasma used for arc discharge is also a hot plasma. In the plasma generating device, the arc discharge between the cathode and the anode makes the flow into the working gas ionization, the output plasma jet, used as plasma jet or plasma jet flame. DC discharge method is used in the device. The polarity of the voltage applied to the electrode is constant in time. The positive potential side is the anode, and the negative potential side is called the cathode. When the electrode is charged to a certain value, the working gas between the electrodes is ionized to form a mixture of positive ions and electrons, which is sprayed out of the generator port at high speed. At this time, the flame has a very high enthalpy, the flame core temperature can reach 10,000 °C~ 20,000 °C, the outer flame part of 0~3000 °C. It is the characteristics of high temperature and high enthalpy of electric arc discharge ionons that are preferred by foreign advanced countries as one of the technologies for harmless treatment of various industrial toxic and harmful wastes, and has been gradually widely used. When using air as the working gas, the continuous working time can reach 30 h, and when using nitrogen as the working gas, the continuous working time can reach more than 100 h, which is stable and reliable, paving the way for the application of arc plasma generator in the experimental device.

At present, the dielectric barrier discharge plasma biomass liquefaction technology has a series of advantages, such as fast reaction speed, mild conditions, simple equipment, stable product properties, low corrosion, but the conversion efficiency is low. Low-temperature plasma biomass refining technology [8] can provide a reaction environment with high density of active particles and high energy, and has some characteristics better than conventional technologies in the biomass refining process, but there are still breakthroughs to be made in key technologies such as energy efficiency of high-voltage power supply, gas–solid efficient mixing and reactor amplification [9].

In view of the above theories and existing technologies, the team determined the research program of plasma biomass straw liquefaction device through research, and carried out in-depth research on the structure of plasma biomass straw liquefaction device. The plasma straw liquefaction unit has a series of advantages such as novel method, high conversion rate and low energy consumption.

4 Plasma Biomass Straw Liquefaction Unit

4.1 Overall Structure of the Device

The device consists of a plasma power supply, a gas mixer, a multifunctional feeder, a plasma conversion reactor, a detection device, a cyclone separator, a condenser, a gas collection device and a big data analysis platform.

For lignocellulose, the adsorbed water can be decomposed at 25~240 °C, and some glucose groups begin to dehydrate. Low molecular weight volatile compounds were produced when the temperature was higher than 240 °C. Therefore, the reaction temperature range is 25~240 °C, as the basis of reactor design. In addition, straw is easy to burn and oxidize, producing volatile compounds with low molecular weight. Products such as light bio-oils may also decompose and produce volatiles under plasma conditions, reducing production. In the reaction process, the straw powder is the stationary phase and the exhaust gas is the mobile phase. The reaction product is adsorbed on the surface of the reaction material. Only by removing the product with mobile phase, increasing the mass transfer rate and making straw powder contact with exhaust gas, can the yield be effectively increased. Therefore, in the discharge reactor, the discharge gas needs to be in full contact with the reaction material to provide an inert gas environment and ensure a certain amount of water vapor flow, and the intake system is designed to quickly remove the reaction products. By controlling the plasma conversion reactor, the solid, liquid and gaseous products pass through the separator and condensation tube successively, and the solid particles or droplets with large inertial centrifugal force are thrown to the outer wall by the rotation of the separator, and the liquid products are collected by the condenser. Figure 1 is the flow chart of the experimental reaction process and the composition of the system experimental device.

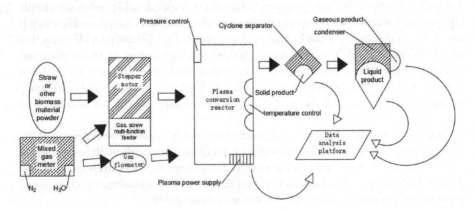

Fig. 1 System process flow chart

4.2 Basic Introduction of Arc Plasma

Arc plasma generator [10] is mainly composed of three parts: a pair of discharge electrodes, a discharge chamber and plasma working gas supply system. The discharge principle of plasma generator is to conduct gas by applying applied electric field or high-frequency induction electric field, which is called gas discharge [11]. Gas discharge is one of the important means to produce plasma. Electrons in a partially ionized gas, accelerated by an applied electric field, collide with neutral molecules and transfer energy from the electric field to the gas. The elastic collision between electrons and neutral molecules results in the increase of molecular kinetic energy, which is manifested as temperature rise. Inelastic collisions, on the other hand, result in excitation (the transition of electrons in a molecule or atom from a lower energy level to a higher energy level), dissociation (the decomposition of molecules into atoms), or ionization (the transition of electrons in the outer layer of a molecule or atom from a bound state to a free one). High-temperature gases transfer energy to the surrounding environment by conduction, convection, and radiation. Under steady conditions, the energy input and loss in a given volume are equal.

4.3 Power System

The data in the Table 1 is the technical function parameter table of system power supply. The plasma source is connected by a three-phase four-wire system with a 380 V ± 10 % AC power supply and a frequency of 50 Hz. The constant current output current range of the power system is DC 40A ~60A, the working voltage is 40 ~130 V, the total output is not less than 7 kVA. Cross flow control is adopted. The internal cooling mode of power supply is air cooling. Four kinds of protection are adopted: short circuit protection, over current protection, water shortage protection, lack of phase protection. The display area is used to display voltage, current, power, running time, startup times, running status and fault code. The main function of the control area is to start, stop, pause, modify parameters, short-range control and remote control, etc.

Table 2 shows the range of output power adjustments for plasma and generator. Cooling water flows for anode and cathode can be obtained from the table. The optimal cooling water flow rate is obtained through the experiment. The anode cooling water flow rate is 1~2 m^3/h, and the cathode cooling water flow rate is 1~2 m^3/h.

Table 1 Power supply technical function parameter table

Output power	Input power	On-load output voltage	Output current	Open circuit voltage
2~6 kW	~380 V	40 V~130 V	30A~60A	170 V

Table 2 Plasma and generator output power adjustment range

	Power range	The resulting arc length is zero	Nitrogen gas flow regulation	Argon gas flow regulation	Gas pressure
When nitrogen is the working gas	3~6 kW	2~3 cm	0.5~0.9 $m^3/h \cdot N$	×	2~3 MPa
Argon is the working gas	2~3 kW	1~2 cm	×	0.5 ~1 $m^3/h \cdot N$	2~3 MPa

4.4 Gas Supply System

The gas supply system is mainly composed of external gas cylinders, gas mixer and gas flowmeter. Argon and water vapor are used as reaction materials in the device. Argon and water vapor are passed through a gas flow meter, which is used to monitor and control the flow of gas into the plasma conversion reactor. The function of the gas mixer is to mix argon gas and water vapor in accordance with the proportion required by the experiment to improve the reaction effect.

4.5 Monitoring System

The monitoring system of the device mainly consists of temperature sensors and pressure sensors. Temperature sensors are evenly distributed in the plasma conversion reactor, which are used to monitor the temperature of different positions in the reactor, so as to ensure the full reaction of biomass straw at the optimal temperature in the conversion reactor. Pressure sensors are used to monitor the gas pressure in the reformer reactor. When the pressure exceeds the normal range, an alarm will be given. The pressure sensor is connected with the pressure reducer. The working principle of the pressure reducer is to depressurize the high-pressure gas in the cylinder for use. The pressure reducer can maintain stable outlet pressure under the condition of changing the input pressure and the input flow. Through monitoring the plasma conversion reactor, the monitoring system controls the inlet rate, feed rate and power of the device to ensure the quality of the product.

4.6 Product Separation System

The product separation system includes cyclone separator, cooler and gas recovery device. After the reaction of straw and other biomass powder in the plasma conversion reactor, the generated products first enter the cyclone separator, which separates the products into three states of solid, liquid and gas. Solid products are mainly carbon black, carbon black has the role of reinforcing, wear-resisting and filling, at the same time can be used as a pigment, but also used in the rubber tire industry, the main role of carbon black for special purposes is coloring, used in ink coatings and other industries. The liquid product obtained by the reaction is separated by fractionation into water, light oil and heavy oil. Light oil collects products with boiling point lower than 100 °C, and heavy oil collects products with boiling point higher than 200 °C [3]. The product obtained is combustible and can be fully used as industrial raw materials after separation, in addition, liquid products also include carbohydrates and lipids that can be reused. Subsequently, it can be separated by targeted catalysis and reused. The remaining gas products mainly consist of carbon monoxide and alkanes, and very little carbon dioxide and nitrogenous gases, some of which can be used as fuel. Through the product separation device, the transformed product generated solid liquid gas three states can be reused environmental protection products [12].

5 Experimental Methods and Analysis of Results

5.1 Experimental Methods

After determining the experimental scheme, the team used TRIZ theory to continuously optimize the existing device in the experiment [13]. Firstly, the TRIZ theory is used to determine the technical contradiction: under the condition of straw conversion, the contradiction occurs between increasing the mechanism and device and increasing the efficiency of the device. The second is to determine the physical contradiction, using the principle of space separation: the conversion device should be high efficiency to meet the transformation demand in the enterprise demand space; converters should be inefficient to meet cost requirements in terms of equipment cost performance space. From the principle of time separation, it is concluded that the conversion device should improve the decomposition efficiency of straw to meet the needs of straw conversion in the time period required by users. The conversion device should reduce the straw decomposition efficiency to meet the strength requirements of the parts during the equipment use period. On the principle of relationship separation: the converter should increase the strength of equipment to meet the demand of performance strength for the use of objects; the converter shall reduce the strength of the equipment to meet the cost requirements for the procurement object. The optimization scheme was proposed by using the contradiction method, and the components of the device and the optimal experimental conditions were selected [14].

5.2 Analysis of Experimental Results

Using the principle of space separation, the technical scheme is put forward: plasma electrolysis technology is used in the conversion process of the conversion device, which achieves high conversion efficiency and low cost. According to the principle of time separation, the straw is decomposed by plasma flame, which greatly improves the reaction efficiency. The principle of relational separation team obtained the plasma reaction converter with cylindrical upper closure device scheme and applied, so that the device takes into account the cost and performance strength. The ionospheric biomass straw liquefaction device shorts the straw treatment process and is easy to operate and treat in situ. The device can be powered by solar energy and easily moved to the end of the field. It can process biomass straw while harvesting crops. By studying the working gas, reactor pressure, reactor temperature, etc., to realize the conversion of biomass straw in plasma, reduce the amount of acid, alkali and water, shorten the process flow.

6 Conclusion

The technical proposal of the team to treat biomass straw through plasma, the thermal efficiency of biomass liquefaction products produced by the developed plasma biomass straw liquefaction unit is 48 %, and the conversion rate of biomass straw is 90.4 %. Device with many innovations, the first is gas discharge and the discharge plasma technology was applied to the biomass straw processing, novel methods, plasma straw biomass liquefaction device adopts many kinds of discharge form, response sufficiently simple and efficient, and mature work can realize the biomass straw on-site processing, energy conservation, environmental protection, in line with the "carbon peak, carbon neutral" development philosophy. Therefore, the technology has good prospects for popularization and application. The team will continue to improve the performance of the product through follow-up experiments.

Acknowledgements This work was financially supported by the National Natural Science Foundation of China (Grant No. 51877024), University Nursing Program for Young Scholars with Creative Talents in Heilongjiang Province (Grant No. 2020205), Natural Science Foundation of Heilongjiang Province, China (Grant No. LH2020E111), 2022 Special Foundation Project of Fundamental Scientific Research Professional Expenses for Undergraduate Universities in Heilongjiang Province, Innovation and Entrepreneurship Training Program for College students of Heilongjiang University of Technology (Grant No. 202111445010, 202111445041), Science and Technology Program Project in Jixi City, Heilongjiang Province, China.

References

1. Wang Q, Gu F (2010) Biomass liquefaction by dielectric barrier discharge plasma. Trans Chin Soc Agric Eng 26(02):290–294
2. Ghaffar SH, Fan M (2013) Structural analysis for lignin characteristics in biomass straw. Biomass Bioenergy 57:264–279
3. Tian S-Q, Zhao R-Y, Chen Z-C (2018) Review of the pretreatment and bioconversion of lignocellulosic biomass from wheat straw materials. Renew Sustain Energy Rev 91:483–489
4. Zhang J-J, Zhang X (2019) The thermochemical conversion of biomass into biofuels. In: Biomass, biopolymer-based materials, and bioenergy. Woodhead Publishing, pp 327–368
5. Lu J-W et al (2022) Elemental migration and transformation during hydrothermal liquefaction of biomass. J Hazard Mater 126961
6. Fu X-K et al (2020) Class straw biomass pretreatment technology research progress. J Chem Ind Shandong 49(01):36–38
7. Liu Y, Wang Y, Wang Y et al. Application of biological treatment technology to the standardization and commercialization of straw feed. J Jilin Anim Sci VetY Med, 201, 42(01):100
8. Fu X-G, Chen HZ (2014) Biomass refining technology with low temperature plasma. Chin J Bioeng 30(05):734–752
9. Wu C, Yan BH, Zhang L, Shuang Y, Jin Y, Cheng Y (2010) Key technology and economic analysis of one-step acetylene process by thermal plasma pyrolysis. J Chem Ind Eng 61(07):1636–1644
10. Du Z-Z (2016) Research on jet flow characteristics of arc heated plasma engine. Nanjing Univ Sci Technol
11. Hao C (2014) Research on plasma flow characteristics under arc heating. Nanjing Univ Sci Technol
12. Hou JM (2019) Study on catalytic pyrolysis of tar during biomass energy utilization. Hebei Univ Sci Technol
13. Zhao S (2021) Brief introduction of TRIZ Theory. J Beijing Pet Manag Cadre Inst 28(03):36
14. Xu X, Su Y, Jia J-X, Sun Y, Zhao Z-N (2021) Optimization design of coke reactivity detector based on TRIZ theory. Guangzhou Chem Ind 49(19):113–114+137

References

Analysis of Dust Deposition in Highland Areas for the Effect of Photovoltaic Modules Performance

Jiahao Wu, Jeilei Tu, Lei Li, Kai Hu, Shouzhe Yu, Hao Wu, Yucen Xie, and Yanyun Yang

Abstract When PV modules are exposed to the outdoors for a long time, the module glass surface is prone to lower module conversion efficiency due to dust deposition, as well as jeopardize the normal operation of the modules. By researching the characteristics of dust deposition in the Yunnan plateau region, the performance of PV modules under plateau climate was analyzed, the transmittance of PV modules under different deposition densities was calculated, the effect of different deposition densities on the performance of PV modules was tested, and the output performance of the modules under actual operating conditions was compared. The dust deposition density on the PV module surface was 0.957 g/m^2 over one month. The density of dust deposition will reach 2.016 g/m^2 in half a month without considering the weather conditions. The irradiance had a significant effect on the transmittance, and the change in transmittance of clean glass at $(550–850)$ and $(850–1050)$ W/m^2 was 4.56 and 0.57%, respectively. The effect of dust deposition on PV modules power and current had a decreasing linear relationship. One month of dust deposition resulted in a 7.161% drop in output power and a 6.618% drop in output current. This has important implications for the selection of suitable dust removal cycles in highland areas and the reduction of operating costs.

Keywords PV modules · Dust deposition · Output performance · Irradiance · Transmittance

1 Introduction

With the development of solar cell technology, solar power will gradually replace fossil energy generation as the main source of supply energy in various fields [1]. The solar cell was a device that directly converts light energy into electricity through

J. Wu · J. Tu (✉) · L. Li · K. Hu · S. Yu · H. Wu · Y. Xie · Y. Yang
Solar Energy Research Institute, Yunnan Normal University, Kunming 650500, China
e-mail: tjl@ynnu.edu.cn

Yunnan Provincial Key Laboratory of Rural Energy Engineering, Yunnan Normal University, Kunming 650500, China

Y. Ma (ed.), *Advanced Theory and Applications of Engineering Systems Under the Framework of Industry 4.0*, https://doi.org/10.1007/978-981-19-9825-6_17

209

the photoelectric or chemical effect. For practical use, the solar cells were often used to generate electricity in the form of photovoltaic (PV) modules. The PV modules exposed to the outdoors could be susceptible to dust deposition due to their surface. The dust particles floating in the air could adhere to the surface of PV modules and affect the transmittance of PV modules, thus affecting the output performance of PV modules [2–5]. Due to the climatic characteristics of high altitude, dryness, and low rainfall, the dust deposition on the surface of PV modules was more serious.

In recent years, domestic and foreign scholars have done a lot of research on dust deposition on component surfaces. Aslan Gholami et al. found that the rate of dust deposition would slowly slow down with experimental time during the two months without rain and reached $6.0986 \, g/m^2$ on the PV modules surface during this period, with a 21.47% reduction in output power [6]. To reduce dust deposition on PV modules, Benjamin Figgis et al. found through field measurements and modeling that by applying floating PV in a desert environment, the desert seaside deposition rate is reduced by up to 60% compared to desert land [7]. Sangchul Oh, through analytical models and Monte Carlo simulations, proposed that dust with finer particle size would lead to a reduction in transmittance, and the dust with larger particle size would have greater fluctuations in its reduction in transmittance and show normal and log-normal distributions, respectively [8]. Also, Arash Sayyah et al. found that the finer dust particle size deposited on the PV modules had a larger specific surface area compared to the larger dust particle size, which resulted in more scattering losses, and the wind speed and dust particle size were found to play a major role in the irradiance attenuation rate and the dust deposition density deposited on the PV modules surface. A similar finding was made by Dirk Goossens et al. [9]. J. K. Kaldellis and M. Kapsali et al. also pointed out that the accumulation of dust and solid particles on the surface of PV modules causes an increase in solar reflection, which would lead to a decrease in solar cell absorption [10]. Letin et al. also explored the effect of dust particles deposited on the surface of PV modules on the performance degradation and showed that the equivalent series–parallel resistance was strongly affected due to the performance degradation of PV modules, which led to a decrease in the fill factor (FF) of PV modules [11]. Ze Wu et al. also proposed a mathematical model to predict the effect of dust on the transmittance of PV modules based on the shape of dust particle size, and the experimental results differed from the model prediction by only 1% [12].

The effect of dust deposition on PV modules performance was also related to regional weather conditions [13]. Adinoyi and Kalogirou et al. compared the effect of dust deposition on PV modules power generation efficiency during the rainy and dry seasons, and the results showed that the long dry season resulted in 20% less power generation and the module efficiency would decrease by 25% due to dust deposition compared to the rainy season, and the module efficiency could decrease by up to 50% if the module surface was not cleaned for six months. Darwish and Kazem et al. also indicate that a very thin layer of dust deposited during the dry season could reduce solar energy conversion efficiency by 40% [14]. Jinxin Chen and Lu et al. also developed a model for predicting dust deposition under natural rainfall cleaning by improving, for example, the inertia factor decay method in the particle swarm algorithm (PSO), which can quickly predict as well as count the amount of PV modules dust deposition and power decay rate during the cycle [15]. Experiments conducted in Surabaya, Indonesia, showed that the higher the percentage of relative humidity in the air, the lower the output power, and when the average relative humidity in the air reaches 76.32%, the module output power would be reduced by more than 40% [13].

Although the above work has been studied in detail for the dust deposition on the surface of PV modules, the dust deposition on the surface of PV modules in highland areas varies in terms of the rate of surface deposition and the impact on the output performance of PV modules due to different weather and geographical conditions. This experiment investigated the effect of dust deposition on the output performance of PV modules in the Yunnan plateau area and the relationship with power and current and found the effect law of deposition time on the decay of irradiance transmittance. This research was significant in removing dust from PV module surfaces in highland areas, guiding the selection of dust removal cycles, and increasing the efficiency of PV power generation.

2 Experimental Protocol

The PV modules used in this test, which lasted for one month, were crystalline silicon solar modules with the parameters shown in Table 1. The test was conducted using the natural deposition method (PV modules were placed in an open outdoor area, and the mounting angle of PV modules and irradiation sensors was set at 25° using a special module holder), and three pieces of 100 mm × 100 mm × 3.2 mm ultra-white suede tempered glass were placed on the bottom of the module surface to measure the density of dust deposition on the module surface and the transmittance of dust deposition, as shown in Fig. 1. A balance and a UV-Vis-NIR spectrophotometer were used to analyze the amount of dust deposition and the transmittance of the glass, respectively. The instrument applied to measure the property values of the PV modules is the portable IV tester: IV-525 W, manufactured by HT, Italy, which carried an irradiation sensor to accurately measure the solar irradiance.

For the calculation of solar irradiance attenuation rate [16], it is mainly calculated by Eqs. (1)–(3):

$$\zeta_{Dust} = \zeta_{Tolal} - \zeta_{Glass} \tag{1}$$

$$\zeta_{Tolal} = 1 - \frac{\left| \frac{\sum_1^n I_{rr-Under}}{n} - I_{rr-Total} \right|}{I_{rr-Total}} \tag{2}$$

$$\zeta_{Glass} = 1 - \frac{\left| \frac{\sum_1^n I'_{rr-Under}}{n} - I_{rr-Total} \right|}{I_{rr-Total}} \tag{3}$$

Table 1 Main parameters of crystalline silicon cell modules

Parameter	Value
Type	Monocrystalline solar module
Pmax	60 W
Vmp	18 V
Imp	3.33 A
Voc	21.6 V
Isc	3.93 A
Dimensions (L × W × H)/mm	680 × 510 × 30
Maximum system voltage	500 V
Standard test	AM 1.5, 1000 W/m², 25 ± 2 °C

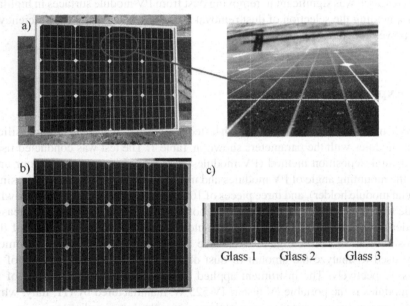

Fig. 1 Collection of dust deposition samples of PV modules: **a** dust deposits on the PV module surface, **b** clean PV modules, **c** glass samples

Among them, ζ_{Dust} (%) is the transmission attenuation rate of dust deposition; ζ_{Total} (%) is the total attenuation rate of solar irradiance when the dust is deposited on the glass; ζ_{Glass} (%) is the transmission attenuation rate of the deposited glass; $I_{rr-Total}$ (W/m²) is the upper solar irradiance test value of the dust deposited glass; $I_{rr-Under}$ (W/m²) is the solar irradiance measurement value of the lower part of the glass in the clean state; $I'_{rr-Under}$ (W/m²) is the solar irradiance test value of the lower part of the glass under the condition of dust deposition; n was the number of locations where the PV modules were taken.

3 Experimental Results and Discussion

3.1 Analysis of Dust Deposition Density Variation

The quality of dust deposition was calculated through periodic weighing of the tempered glass placed on the surface of the PV module, and then the density of dust deposition was calculated from the area of the glass. The mass of dust was measured by an electronic balance (the precision was 0.0001 g), and the curve of dust deposition density versus time was shown in Fig. 2.

From the figure, the density of dust deposition on the surface of the PV modules fluctuated from high to low during the one-month experimental period, which was mainly caused by the weather, as shown in Table 2. From the trend of the curve, which could be seen that the slope of the deposition density gradually became slower on the 8th day of the experimental period, which was mainly attributed to the windy weather, although it did not rain at the experimental site, the air was more humid. The dust particles deposited on the PV modules surface would generate capillary force with the glass surface and the vertical adsorption force would also increase, and the adhesion force would be inversely proportional to the particle size, and the adhesion force of small particle size dust would be stronger than the peeling force of wind. The reason for the decrease in deposition density in the figure was that the dust on the surface of the modules was run off with the rain for a long time, which made the deposition density lower. Although cleaned by rain twice, the dust still existed on the module surface, and the erosion of rain made the humidity on the PV modules surface rise, and under the influence of humidity, the dust particles would have higher adhesion to the glass. Although rainfall could play a certain role in cleaning the modules, as dust is deposited for a long time, a cementation process is created between dust and dirt, and some dust particles adhere to the surface of the modules, resulting in rainwater does not do a thorough cleaning of the PV modules.

Fig. 2 Variation curve of dust deposition density with time

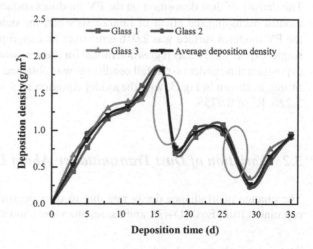

Table 2 Weather changes at the experimental site

Date (d)	Highest temperature (°C)	Lowest temperature (°C)	Weather	Wind speed and direction
8	23	8	Cloudy	9.6 m/s Southeast
9	25	11	Cloudy/Wind	7.3 m/s Southwest
17	22	9	Moderate rain	4.9 m/s Southwest
18	23	9	Light rain, shower	4.2 m/s Southwest
25	22	11	Shower	6.9 m/s Southwest
26	24	10	Light rain	5.1 m/s Southwest
27	24	11	Light rain	3.9 m/s Southwest

Weather data was derived from on-site tests

The weather was the principal cause of the variation in dust deposition density. Although windy weather could remove part of the larger dust particles deposited on the module surface, which also increased the amount of dust floating into the air, eventually leading to faster dust deposition on the module surface. Meanwhile, the weight of the deposited glass was weighed by electronic balance, and the dust weight was found to drop a small amount in one or two days after the rainfall, which was mainly because the air humidity was higher after the rainfall and there were fewer dust particles in the air, and there was more windy weather in Kunming, so the density of dust deposition on the PV modules surface decreased a small amount (the first deposition density decreased faster mainly because of the continuous rain). The density of dust deposition on the PV modules surface was 0.957 g/m^2 over one month; excluding the effect of rainfall weather, the density of dust deposition on the PV modules surface was 2.016 g/m^2 over an experimental period of 16 days. A single-exponential decay regression model for the variation of deposition density with deposition time under no rainfall conditions was obtained employing nonlinear curve fitting, as shown in Fig. 3, with the model equation is: $y = -3.177 * \exp(-\frac{x}{18.796}) + 3.226$, R^2 of 0.9755.

3.2 Variation of Dust Transmittance About Deposition Time

The change of irradiance decay rate due to dust deposition in each test would be calculated from Eqs. (1)–(3), and the results were shown in Fig. 4.

Fig. 3 Fitting curve of deposition density with deposition time

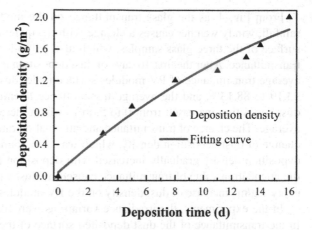

Fig. 4 Deposition time variation in the relation of **a** transmittance, **b** transmittance attenuation value

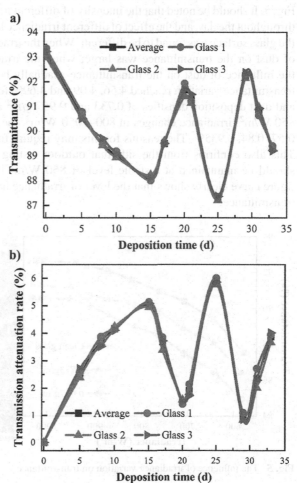

From Fig. 4, as the glass transmittance on the module surface increases with rainfall, windy weather causes a change with the quantity of dust deposited on the surface of the three glass samples, which also led to a change in the glass surface transmittance. After the first 16 days of dust deposition in the absence of rainfall, the average transmittance of PV modules surface under dust coverage decreased from 93.19 to 88.13%, and the average transmittance attenuation value resulting from dust deposition increased from 0 to 5.06%, which increased by 0.32% per day on average. The change of transmittance and attenuation rate was mainly caused by the change of dust deposition density, which was also confirmed by the result that the deposition density gradually increased with deposition time in Fig. 2. Meanwhile, the attenuation value of transmittance caused by dust would increase by 2.51% for every 1 g/m^2 increase of dust density on the PV modules surface.

In the experiments, the irradiance variations were also found to cause changes in the transmittance of the dust deposited surface of the PV modules, as shown in Fig. 5. It should be noted that the intensity of different irradiance levels was recorded throughout the day, and the effect of different irradiance levels on the transmittance of the glass surface was found to be different. When the irradiance was low, the influence of dust on the transmittance was large; when the irradiance gradually increased, the influence of dust on the transmittance gradually became smaller. The average transmittance variation reached 4.56, 4.09 and 4.69% for glasses with surface clean and dust deposition densities of 0.733 and 0.957 g/m^2 at irradiance ranges of 550–850 W/m^2; irradiance changes in 800–1050 W/m^2, the transmittance changes only 0.57, 0.84, 0.935%. The reasons for this may require further research in the future. This also confirms from the side that outdoor testing of PV modules, irradiance should be maintained at a stable level of 850 W/m^2 or more. The change of the figure curve clearly shows that the level of irradiance had a significant effect on the transmittance.

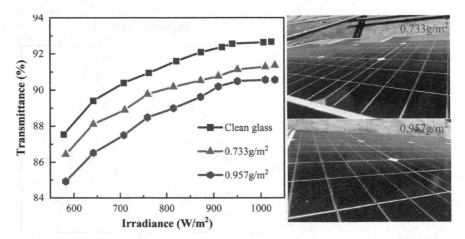

Fig. 5 The influence of irradiance variation on transmittance

3.3   Effect of Dust Deposition on PV Modules Power and Current

The dust deposition on PV modules surface would cause the increase of modules to surface temperature and decrease of transmittance, which would lead to the weakening of PV effect, so the output power, output current, etc. would decrease with the increase of deposition time. As shown in Fig. 6a, c, the density of dust deposition on the surface of the modules would gradually increase with time, and the power and current of the modules would decrease to different degrees. During the 16 days of deposition without rainfall, the dust deposition density reached 2.016 g/m² and the output power decreased by 9.479% and the output current decreased by 9.191%, averaging 0.593 and 0.575% per day, respectively. While the dust deposition density reached 0.957 g/m² in one month, the output power of PV modules decreased by 7.161% and the output current decreased by 6.985%.

A linear fit of output power versus deposition time derived from the first 16 days yielded a fitting equation of y = 48.54814 − 0.28958x, with a coefficient of determination R² of 0.98758, and a fitting equation of output current versus deposition

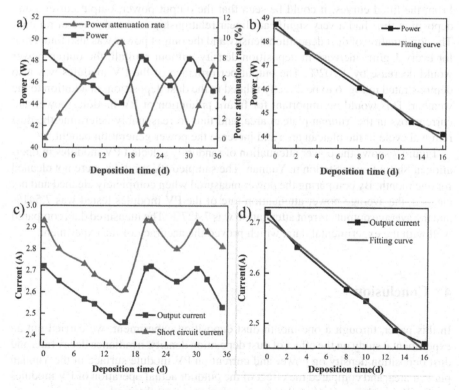

Fig. 6 Effect of deposition time on PV modules performance: **a** effect on power, **b** power fitting curve, **c** effect on current, **d** fitting curve to current

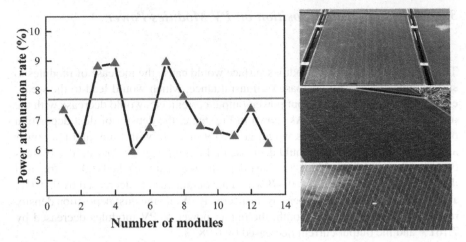

Fig. 7 Effect of dust on module power under outdoor conditions

time off y = 2.71132 − 0.0179x, with a coefficient of determination R^2 of 0.98754. From the fitted curves, it could be seen that the output power, output current, and deposition time had a very significant linear relationship in the absence of rainfall. The combination of dust deposition density and the output power was also found that for every 1 g/m^2 increase in deposition density without rainfall, the output power would decrease by 4.702%. The output performance of the PV modules was also demonstrated in Fig. 6 to be directly related to the dust deposition, in addition to the weather. This would be important for future prediction of PV modules power and current loss in the Yunnan plateau area, and thus to reasonably determine the dust removal cycle in the plateau area and increase the power generation benefit.

Figure 7 shows the power attenuation of randomly selected PV modules in operation at Shilin Power Station in Yunnan. The sampled PV modules were not cleaned for one month. By comparing the power measured when completely cleaned and not cleaned, the average power attenuation rate of the PV modules tested was 7.541% and the average output current attenuation was 7.427%. The measured data compared well with the experimental data, which proved the accuracy of the experiment.

4 Conclusion

In this paper, through a one-month dust deposition experiment, we carried out an experimental study on the effect of dust deposition density, irradiance decay rate, and dust deposition density on power and current on PV module surface in the Yunnan plateau area, and compared the effect of the outdoor actual operation of PV modules on dust deposition, and the following conclusions were obtained.

(1) A single-exponential attenuation regression model was obtained for the variation of sediment density with sedimentation time under no rainfall conditions. As the outdoor exposure time increases, the variation of dust deposition density on the PV modules surface would be leveled off, and the dust deposition density was 0.957 g/m^2 during the one-month experimental period. Excluding the effect of rainfall, the dust deposition density reached 2.016 g/m^2 over the 16-day experimental period.

(2) The transmittance of PV modules was influenced by the density of dust deposition. The linear relationship between dust deposition density and transmittance attenuation was 2.51%/g/m^2.

(3) The output performance of PV modules was linearly and negatively correlated with the deposition time. During the 16 days without rainfall, the output power of PV modules decreased by 9.479% and the output current decreased by 9.191%; under one month's natural conditions, the output power of PV modules decreased by 7.161% and the output current decreased by 6.985%.

(4) For every 1 g/m^2 increase in dust deposition density, the PV modules output power would be reduced by 4.702%.

(5) Although natural rainfall could clean some of the dusts, it still needs to be treated by humans or machines for the dust and dirt that produce cementation on the surface of PV modules.

Acknowledgements This work was supported by National Nature Science Foundation of China (Grant Number 61664010).

References

1. Lehtola T, Zahedi A (2019) Solar energy and wind power supply supported by storage technology: a review. Sustain Energy Technol Assess 35:25–31
2. Mehmood U, Al-Ahmed A, Al-Sulaiman FA et al (2017) Effect of temperature on the photovoltaic performance and stability of solid-state dye-sensitized solar cells: a review. Renew Sustain Energy Rev 79:946–959
3. Abderrezek M, Fathi M (2017) Experimental study of the dust effect on photovoltaic panels' energy yield. Sol Energy 142:308–320
4. Bagheri-Bodaghabadi M, Jafari M (2022) The dust deposition model (DDM): an empirical model for monitoring dust deposition using meteorological data over the Isfahan province in central Iran. CATENA 211:105952
5. Zhao W, Lv Y, Zhou Q et al (2021) Investigation on particle deposition criterion and dust accumulation impact on solar PV module performance. Energy 233:121240
6. Gholami A, Khazaee I, Eslami S et al (2018) Experimental investigation of dust deposition effects on photo-voltaic output performance. Sol Energy 159:346–352
7. Fountoukis C, Figgis B, Ackermann L et al (2018) Effects of atmospheric dust deposition on solar PV energy production in a desert environment. Sol Energy 164:94–100
8. Oh S (2019) Analytic and Monte-Carlo studies of the effect of dust accumulation on photovoltaics. Sol Energy 188:1243–1247

9. Goossens D, Lundholm R, Goverde H et al (2019) Effect of soiling on wind-induced cooling of photovoltaic modules and consequences for electrical performance. Sustain Energy Technol Assess 34:116–125
10. Kaldellis JK, Kapsali M, Kavadias KA (2014) Temperature and wind speed impact on the efficiency of PV installations. Experience obtained from outdoor measurements in Greece. Renew Energy 66:612–624
11. Letin VA, Nadiradze AB, Novikov LS et al (2005) Analysis of solid microparticle influence on spacecraft solar arrays. In: Conference record of the thirty-first IEEE photovoltaic specialists conference, vol 2005, pp 862–865
12. Wu Z, Yan S, Ming T et al (2021) Analysis and modeling of dust accumulation-composed spherical and cubic particles on PV module relative transmittance. Sustain Energy Technol Assess 44
13. Khanam S, Meraj M, Azhar M et al (2021) Comparative performance analysis of photovoltaic modules of different materials for four different climatic zone of India. Urban Clim 39
14. Hamou S, Zine S, Abdellah R (2014) Efficiency of PV module under real working conditions. Technologies and materials for renewable energy, environment and sustainability (TMREES14-EUMISD) 50:553–558
15. Jinxin C, Guobing P, Jing O et al (2021) Study on prediction algorithm of dustfall on PV modules under natural rainfall. Acta Energiae Solaris Sinica 42(02):431–437
16. Shengjie W, Rui T, Xiao G et al (2019) Dust accumulation characteristics and transmission attenuation law of photovoltaic modules. Trans Chin Soc Agric Eng 35(22):242–250

Research on Efficiency Improvement Scheme of Multi in One Powertrain

Xuhui Liao, Xuling Wu, and Jianming Yao

Abstract Multi in one powertrain system is the main development trend of large automobile manufacturers at home and abroad. It has the characteristics of high integration and lightweight. In order to improve the efficiency, based on a mass-produced vehicle, taking SiC as the power device, this paper adopts the linear frequency conversion method to reduce the motor noise, reduce the IGBT switching loss and improve the motor efficiency; The maximum torque current ratio control is selected to optimize the connection mode of the inner and outer windings of the motor, reduce the copper consumption and improve the efficiency of the motor. Experiments are carried out to verify the correctness and effectiveness of the theory and optimization method, the results show that the motor noise is significantly reduced under different working conditions, the motor efficiency is increased by 2 % on average, and the area surrounded by the motor high efficiency area is larger.

Keywords All-in-one electric drive assembly · IGBT switch loss · Motor loss · SiC · High efficiency

1 Introduction

In recent years, the problem of global warming has become increasingly prominent, in which internal combustion engine vehicles produce a large amount of exhaust during driving, which accelerates global warming. Therefore, China has increased its support for the development of new energy vehicles. Pure electric vehicles have been greatly welcomed in China. They have the characteristics of high efficiency, low pollution and low noise, making them the innovation hotspot and development direction of the current automobile industry [1].

X. Liao · X. Wu (✉) · J. Yao
Shanghai Dianji University, Shanghai, China
e-mail: wuxuling@saicmotor.com

J. Yao
e-mail: yaojianming@saicmotor.com

© The Author(s), under exclusive license to Springer Nature Singapore Pte Ltd. 2023   221
Y. Ma (ed.), *Advanced Theory and Applications of Engineering Systems Under the Framework of Industry 4.0*, https://doi.org/10.1007/978-981-19-9825-6_18

At the same time, considering the factors such as cost and demand, reducing cost, improving power density and improving efficiency through more reasonable design has attracted the attention of scholars and major OEMs for a long time. Therefore, based on a multi in one powertrain electric vehicle of a company, this paper optimizes the IGBT switching frequency, reduces noise and improves the efficiency of electric drive system; Optimize the connection mode of motor winding, reduce motor loss and improve motor efficiency, and improve the endurance of the whole vehicle.

2   Material and Methods

The total efficiency transmission of multi in one power is shown in Fig. 1 in which PDU is the high-voltage distribution box, MCU is the motor controller, M is the motor and t is the reducer. This paper mainly studies the ways and methods to improve the efficiency of motor controller inverter and motor in system.

The improvement of converter efficiency mainly includes using different power devices and optimizing the switching frequency of power devices. Some studies show that SiC diode, as a reverse diode, can improve the efficiency of inverter [2]. A random space vector modulation technology is adopted to optimize the switching frequency of power devices and reduce the switching loss [3]. This paper adopts the design of three-level inverter, combined with SiC IGBT device, and reduces the loss by optimizing the switching frequency of IGBT.

The improvement of motor efficiency mainly includes two aspects: Optimizing motor control strategy and reducing motor loss. The optimized motor control strategies mainly include the maximum torque current ratio control method, the optimization method based on loss model and the minimum input power strategy method based on search method [4–7]. Due to the high modeling requirements of the loss model method and the difficulty of calculation of the minimum power strategy based

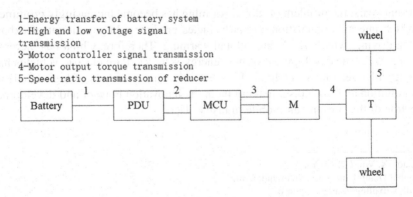

Fig. 1 Transmission of total effectiveness rate of multi in one power

on the search method, the maximum torque current ratio control (MTPA) can minimize the copper consumption and meet the production demand. This paper takes the flat wire motor as the research object, adopts the maximum torque current ratio control, and solves the proximity effect to reduce the motor loss.

2.1 Inverter Efficiency Improvement

Material Selection. IGBT materials of inverter mainly include silicon (SI), germanium (GE), gallium (GaAs), phosphating solution (INP) and gallium nitride (GAN). In recent years, a new generation of semiconductor material silicon carbide (SIC) has the characteristics of high withstand voltage, high electron mobility and strong heat resistance. It has been widely used in motor controller in recent years.

It can be seen from Table 1 that the IGBT made of SiC has a great improvement in withstand voltage and maximum working temperature compared with the IGBT made of GaAs and Si. The system efficiency difference curve of SiC device under 600 V voltage and Si device under 400 V voltage is shown in Fig. 2 It can be seen that SiC power device has significant advantages over Si in improving system efficiency.

Material Selection. The switching loss calculation methods of IGBT module include loss calculation method based on physical method and loss calculation method based on mathematical method. Based on the physical method, the IGBT switching waveform is obtained and the loss is calculated by establishing the relevant physical model; Mathematical method is to model through mathematical method to obtain numerical calculation loss. At present, there are three methods to calculate loss based on data manual, mathematical model and waveform fitting [9]. Because the data manual method is simple and convenient to directly use the data in the data table, but it has the problem of low accuracy, and the mathematical model method is too cumbersome. In this paper, the on and off losses are calculated by waveform fitting method, and the accuracy meets the requirements of this study [10].

IGBT loss includes conduction loss and switching loss, as shown in the following formula (1):

Table 1 Semiconductor material parameters of common power electronic devices [8]

Material	eV	Electron mobility (cm^2/Vs)	breakdown voltage (MV/cm)	Thermal conductivity	Maximum operating temperature (°C)
SiC	3.2	1000	2.8	4.9	600
GaN	3.42	2000	3.3	1.3	800
GaAs	1.42	8500	0.4	0.5	350
Si	1.12	600	0.4	1.5	400

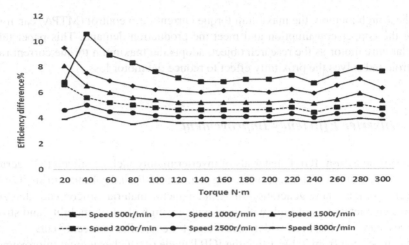

Fig. 2 Efficiency difference curve of power device SiC and SI

$$\begin{cases} P_Q = P_{cond} + P_{SW} \\ P_{cond} = \dfrac{1}{2\pi} \displaystyle\int_0^{\Pi} V_{CE}t \cdot i_C t \cdot \tau t dt \\ P_{sw} = \dfrac{1}{\pi} \displaystyle\sum_{N=1}^{f} E_{(on)} + E_{(off)} \end{cases} \qquad (1)$$

where pcond is the conduction loss, PSW is the switching loss, IC (T) is the function of current and time, and VCE (T) is the saturation voltage drop (terminal voltage), τ (t) Is a function of duty cycle time. Where the switching frequency (carrier frequency) is f, a single IGBT module shall be turned on and off f times in total in unit time, and E (on) and E (off) are the loss of primary energy of IGBT on and off respectively.

It can be seen from the above formula that reducing the switching loss of IGBT can be realized by reducing the switching frequency. In this paper, Infineon IGBT simulation software is used to calculate the power loss and efficiency of each device under a given static load. The specific parameters of DC bus voltage, motor phase current, PWM (pulse width modulation) frequency, modulation strategy, radiator parameters and reference temperature are shown in Table 2.

Adjust the switching frequency to 10 kHz, 8 kHz and 6 kHz respectively. The simulation data are shown in Table 3.

It can be seen from Table 3 that reducing IGBT switching frequency can improve the efficiency of motor controller, with an average increase of 0.3 %.

Based on the actual test of a motor controller at the specified speed and torque, the following data are obtained, as shown in Figs. 3, 4, 5 below:

It can be seen from the above figure that in the full speed range, with the decrease of IGBT switching frequency, the efficiency of motor controller increases by 0.3 % on average, and the simulation is basically consistent with the measured value.

Table 2 IGBT simulation parameters

Type	Value
System frequency	50 Hz
PWM frequency	6 K/8 K/10 kHz
Modulation strategy	Space vector PWM
DC Bus voltage	400 V
Motor phase current	50 A
Power factor	0.8
Thermal resistance	0.1 K/W
Reference temperature	60 °C

Table 3 loss and inverter efficiency under different IGBT carrier frequencies

Switching frequency	IGBT LOSS (W)	Inverter efficiency (%)
6 kHz	358.2	97.04
8 kHz	397.8	96.74
10 kHz	437.4	96.43

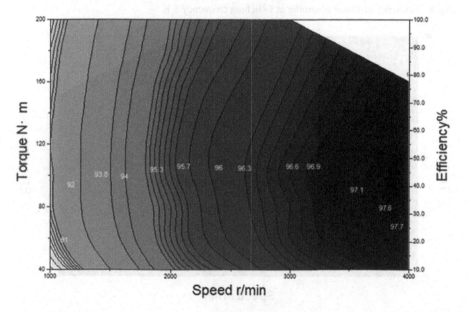

Fig. 3 Efficiency of motor controller at switching frequency 6 K

Reasonable Selection of IGBT Switching Frequency. There are a large number of high-order harmonic components in the current of PWM inverter. The magnetic field generated by fundamental frequency voltage interacts with the spatial magnetic

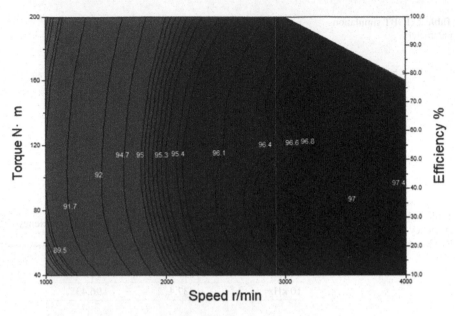

Fig. 4 Efficiency of motor controller at switching frequency 8 K

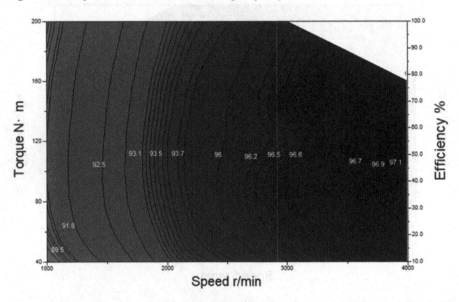

Fig. 5 Efficiency of motor controller at switching frequency 10 K

field generated by high-order harmonic to produce magnetic field force. The electromagnetic noise of motor is the radial pulsation caused by the joint action of electromagnetic force in the air gap between stator and rotor, which makes the motor stator vibrate and produce noise [11]. Taking space vector PWM inverter as an example, the output voltage and current waveform are close to sine wave. Because the carrier is used to modulate the sine signal, the harmonic component related to the carrier is bound to be generated. The harmonic voltage is concentrated at the frequency shown in the following formula.

$$U_{fs} = u_m \cdot [2\pi f_2(t)] \tag{2}$$

FS is the carrier frequency, um is the output voltage amplitude, and UFS is the output voltage.

It can be seen from the above formula that if the carrier frequency changes according to a certain law, the harmonic voltage will no longer be concentrated at a specific frequency, but will be distributed within the frequency time function range of FS (T), and the electromagnetic noise caused by harmonic current will also be distributed within the range of FS (T). In order to reduce the noise of a specific frequency, the variation range of carrier frequency should be as large as possible [11].

In order to reduce the noise at a specific frequency, this paper adopts the linear frequency conversion method. The linear frequency conversion method adopts the linear transformation of switching frequency, adopts different frequencies in different speed segments, and reduces the motor noise on the premise that the motor efficiency remains stable. A large number of tests show that the linear frequency conversion method can meet the requirements. Figure 6 shows the linear frequency conversion method.

In the Actual Test. The efficiency of each speed section is improved when different switching frequencies are adoptedn order to study the noise level and motor efficiency under different switching frequencies, three noise sensors are arranged in the front compartment and one noise sensor is arranged at the right ear of the driver, as shown in Figs. 7, 8, and the motor parameters are shown in Table 4. The vibration bandwidth is set to 12800 Hz, the noise bandwidth is set to 25600 Hz, the resolution is 1 Hz, and the order bandwidth is 1order. Under wot (500–7000r/min) and reg (5000–500r/min) working conditions, different switching frequency control strategies are adopted, as shown in Table 5 below. Figures 9, 10 are the noise comparison diagrams of different control strategies under the two working conditions. Figure 11shows the difference of motor efficiency when the motor adopts linear frequency conversion compared with fixed frequency base (9 kHz fixed frequency).

In the whole vehicle test, the difference of vibration and noise under reg working conditions under different carrier frequency strategies is small. Under wot working conditions, the performance of linear frequency conversion strategy is better than that of fixed frequency strategy. According to Fig. 11, the motor efficiency increases by 1.5 % in low speed section, 2 % in transfer speed section and 2.5 % in high speed section.

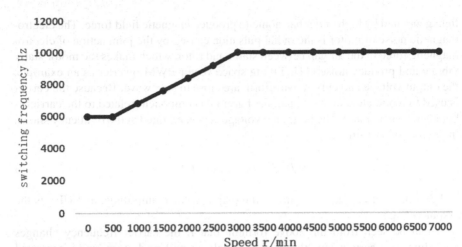

Fig. 6 Linear frequency conversion method

Fig. 7 Motor front compartment noise test

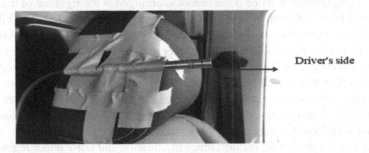

Fig. 8 Driver noise test

Table 4 Motor parameters

Type	Permanent magnet synchronous motor
Number of pole slots	8-stage 48 slot
peak power	70 kW
Peak speed	10000 r/min
Peak torque	200 N·m
Transmission speed ratio	27/79,21/76

Table 5 Control strategies for different switching frequencies

Type	Content
Strategy 1	6 kHz fixed frequency is adopted for 500-3000r/min, and 10 kHz fixed frequency is adopted for 3000r/min and above
Strategy 2	8 kHz fixed frequency is adopted for 500-3000r/min, and 10 kHz fixed frequency is adopted for 3000r/min and above
Strategy 3	10 kHz fixed frequency is adopted in the whole speed section
Strategy 4	Linear frequency conversion
Strategy 5	9 kHz fixed frequency is adopted in the whole speed section

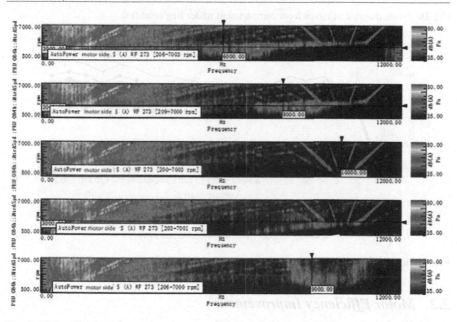

Fig. 9 Noise diagram of motor front compartment under Wot condition

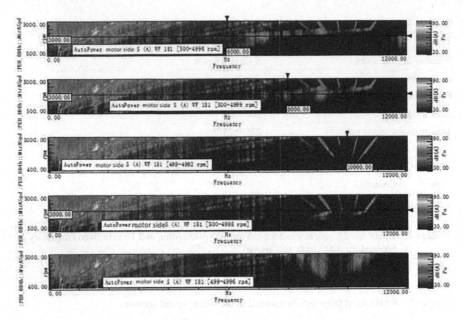

Fig. 10 Noise diagram of motor front compartment under Reg condition

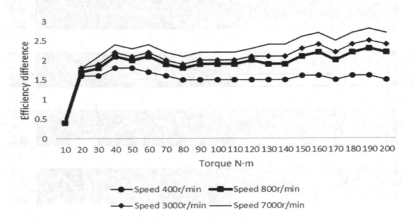

Fig. 11 Motor efficiency difference

2.2 Motor Efficiency Improvement

Motor Control Strategy. The MTPA control strategy is shown in the following Eqs. 3 and 4. When the specified torque is output, the MTPA control strategy is adopted to minimize the amplitude of stator current and reduce the motor loss.

$$\begin{cases} T_e = \dfrac{3}{2} \cdot n_p [\psi_f \cdot i_q + (L_d - L_q) \cdot i_d \cdot i_q] \\ i_s = \sqrt{i_d^2 + i_q^2} \\ H = \sqrt{i_d^2 + i_q^2} + \lambda \left\{ T_e - \dfrac{3}{2} \cdot n_p [\psi_f + (L_d - L_q) \cdot i_d] \cdot i_q \right\} \end{cases} \quad (3)$$

$$\begin{cases} i_d = -\dfrac{\psi_f}{2(L_d - L_q)} + \sqrt{\dfrac{\psi_f^2}{4(L_q - L_d)^2} + i_q^2} \\ i_q = \dfrac{\left(\dfrac{8T_e\psi_f}{3n_p}\right) + \sqrt{\left(\dfrac{8T_e\psi_f}{3n_p}\right)^2 - 4\left[\psi_f^2 - 4(L_d - L_q)^2\right] \cdot \left[\left(\dfrac{4T_e}{3n_p}\right)^2 - \psi_f^2\right]}}{2\left[\psi_f^2 - 4(L_d - L_q)^2\right]} \end{cases} \quad (4)$$

where T_e is the torque, n_P is the speed, Ψ_f is the flux linkage, i_s is the amplitude of stator current, h is the auxiliary function, λ Is the coefficient, L_d and L_q are the inductance of quadrature axis and direct axis, and I_q and I_d are the current of quadrature axis and direct axis.

The principle flow chart of MTPA is shown in Fig. 12 below:

Motor Loss Analysis. This paper takes the flat wire motor as the research object. The flat wire motor changes the traditional motor copper wire from round wire to flat wire. It has the advantages of high slot filling rate, batch generation, strong

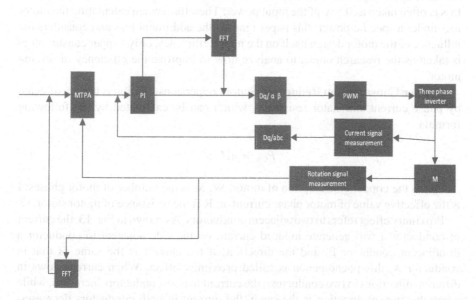

Fig. 12 Flow chart of maximum torque current ratio control principle

heat dissipation capacity, stable structure and high power density. It is the main development direction of electric vehicle drive motor in the future.

The driving motor mainly has proximity effect and skin effect, which affect the motor loss. Skin effect has little effect on motor loss. Only the loss caused by proximity effect is discussed. In order to reduce this loss, the connection mode of inner and outer windings is changed to achieve the purpose.

The efficiency of the motor can be defined as the ratio of the output and input power of the motor, and the calculation formula is shown in formula (5):

$$\eta = \frac{P_2}{P_1} = \frac{P_1 - \sum P}{P_1} \tag{5}$$

P_1 is the input power, P_2 is the output power, and $\sum P$ is the total loss of the motor. According to the above formula, the motor efficiency can be improved by reducing the motor loss.

The total loss of the motor includes the copper loss, iron loss and additional loss of the motor, as shown in the following formula:

$$P_{\text{total}} = P_{Cu} + P_{Fe} + P_{\text{add}} \tag{6}$$

where: p_{total} is the total loss of the motor, P_{Cu} is the copper loss of the motor, P_{Fe} is the iron loss of the motor, and P_{add} is the additional loss of the motor. The motor loss is mainly composed of copper loss, iron loss, mechanical loss and additional loss. For the same operating condition, the mechanical loss can be ignored, and the additional loss is often taken as 0.5 % of the input power. Therefore, when calculating the motor loss under a specific power, this paper ignores the additional loss and considers the influence of the motor design itself on the motor efficiency, only copper consumption is taken as the research object to analyze how to improve the efficiency of driving motor.

Copper Consumption Reduction. Stator copper consumption is mainly affected by phase current and stator resistance, which can be calculated by the following formula:

$$P_{Cu} = mI^2 R \tag{7}$$

P_{CU} is the copper consumption of motor, W; M is the number of motor phases; I is the effective value of motor phase current, a; R is the resistance of motor stator, Ω.

Proximity effect refers to two adjacent conductors. As shown in Fig. 13, the current of conductor a will generate induced current on the side adjacent to conductor a in adjacent conductor B, and the direction of the current is the same as that in conductor A. this phenomenon is called proximity effect. When current flows in different directions in two conductors, the current in both conductors increases, while when the current direction is the same, the current in both conductors decreases. According to this principle, in order to reduce the motor loss, the inner and outer windings of the double-layer motor reduce the current flowing through the winding

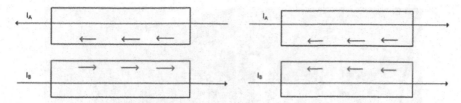

Fig. 13 Schematic diagram of proximity effect

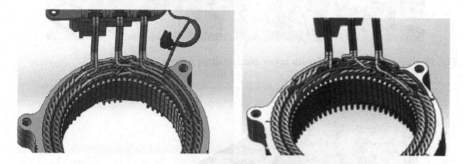

Fig. 14 Inner and outer windings in parallel—inner and outer windings in series

by optimizing the wiring mode to make the inner and outer current directions the same, so as to reduce the copper consumption and improve the motor efficiency [12].

In order to solve the proximity effect between the windings of the double-layer motor, the connection mode of the winding bridge line is changed from parallel to series. As shown in Fig. 14, the inner and outer windings pass through the current in the same direction, the problem of the proximity effect of the inner and outer windings is solved, and the motor efficiency is improved.

The bench test of the motor is shown in Fig. 15 below, and the comparison of the measured efficiency is shown in Figs. 16, 17. It can be seen from the figure that the maximum efficiency of the electric drive system after the change of the winding bridge line has reached 95.8 %, the efficiency of the full speed section has increased by 1 % on average, and the area surrounded by the same efficiency contour is larger.

3 Conclusion

As an important part of electric vehicle, the development trend of powertrain system mainly includes high integration, lightweight, high power density and intelligence. According to the characteristics of high integration and high power density of the multi in one powertrain system, this paper analyzes and calculates the IGBT loss and motor loss respectively. The linear frequency conversion method is adopted to select the IGBT switching frequency compared with the fixed frequency method. Under

Fig. 15 Bench test diagram of multi in one electric drive system

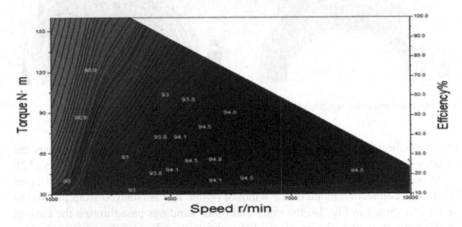

Fig. 16 Efficiency of electric drive system before winding optimization

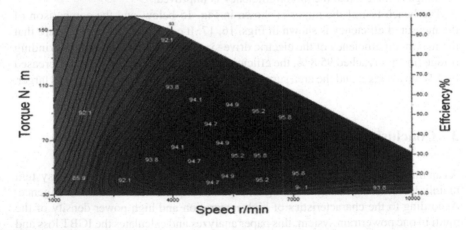

Fig. 17 Efficiency of electric drive system after winding optimization

the condition of reducing the electromagnetic noise of the motor, the efficiency of the motor controller is increased by 2 % on average, the design of the inner and outer windings of the motor is optimized, the proximity effect is solved, and the efficiency of the electric drive system is increased by 1 %. The analysis method in this paper has certain correctness and effectiveness, which improves a certain reference value for the in-depth study of the efficiency improvement of electric drive assembly system.

References

1. Yin Yanqiu, Zhang Junshen (2021) Technical status and development trend of pure electric vehicle drive system Internal combustion engines and accessories, (17): 215–217
2. Mai Yubing, Xie Xinrong (2021) Research progress of the third generation semiconductor material silicon carbide (SIC) Guangdong chemical industry, 48(09): 151–152 + 155
3. Ma Fengmin, Wu Zhengguo, Zhang Zhiqiang (2006) Random pulse width modulation technology based on small switching loss Motor and control applications, (11): 18-21
4. Xu Shufen (2014) Research on efficiency optimization strategy of permanent magnet synchronous motor for electric vehicle Chang'an University
5. Guo Qingding, Chen Qifei, Liu Chunfang (2008) Maximum torque current ratio control method for efficiency optimization of permanent magnet synchronous motor. J Shenyang Univ Technol, (01): 1-5
6. Lin Li, Tang Hongwei, Qiu Xiongyou, Wang Yueqiu (2012) Linearization loss minimization control of vehicle built-in permanent magnet synchronous motor Electric drive, 42(01): 35-39. https://doi.org/10.19457/j.1001-2095.2012.01.008
7. Junggi L, Kwanghee N, Seoho C (2009) Loss minimizing control of PMSM with the use of polynomial approximations. IEEE Transations Power Electron 24(4):1071–1082
8. Suzhi morning Research on IGBT heat dissipation system of high power electronic components North China Electric Power University (Beijing), 2021 https://doi.org/10.27140/d.cnki.ghbbu. Two thousand and twenty-one point zero zero zero two eight six
9. Li Zhigang, Mei Shuang, Wang Shaojie, Yao Fang (2016) Summary of IGBT module switching loss calculation methods Application of electronic technology, 42(01): 10–14 + 18. https://doi.org/10.16157/j.issn. 0258–7998.2016.01.001
10. Zhang Mingyuan, Shen Jianqing, Li Weichao, Geng Shiguang, Tong Zhengjun (2009) A fast IGBT loss calculation method Marine power technology, 29(01): 33-36
11. Zhu Mingxiu, Kong fanhong, Xu Zhexiong (2008) Research and Simulation of random switching frequency PWM (rfpwm) inverter Automation technology and application, (06): 50–52 + 39
12. Zhang Guoquan, Xu yuelang (2010) Study on proximity effect of coils in switching power supply China high tech enterprise, (04): 9–11. https://doi.org/10.13535/j.cnki. 11-4406/n.2010.04.005.

Printed in the United States
by Baker & Taylor Publisher Services